{ A.I }
Beyond Human
Journey Towards A.I World

By: Mr. Deepak Dinesh Kapadnis

EDITION 2023 WORLDWIDE

Title :   Beyond Human - Journey Towards A.I World

Author's Name :   Mr. Deepak Dinesh Kapadnis

Published By :   Self Published

Edition Details :   2023

Copyright © Deepak Dinesh Kapadnis

//SUMMARY

Artificial intelligence, or AI, refers to the capability of a computer or machine to mimic or pretend mortal intelligence and actions. This can include tasks similar as literacy, problem-working, decision- timber, language restatement, and more. There are different types of AI, including narrow or weak AI, which is designed for a specific task, and general or strong AI, which is designed to be suitable to perform any intellectual task that a human can.

AI is frequently achieved through the use of machine literacy algorithms, which allow a machine to ameliorate its performance on a task over time by learning from data and once guests . Machine literacy can be supervised, where the machine is handed with labeled data and a set of rules to follow, or unsupervised, where the machine is given a set of data and must find patterns and connections within it on its own. AI has the implication to revise numerous diligence and make tasks more effective and accurate. It's formerly being used in a variety of fields, similar as healthcare, finance, transportation, and client service. Still, the development and use of AI also raises ethical and societal enterprises, including issues of bias, job relegation, and the eventuality for abuse.

//CONTENT

1] Introduction to Artificial Intelligence   **6 - 10**
    Definition of AI
    History of AI
    Types of AI

2] Machine Learning   **11 - 15**
    Definition of machine learning
    Supervised learning
    Unsupervised learning
    Reinforcement learning

3] Natural Language Processing   **16 - 25**
    Definition of NLP
    Text classification
    Language translation
    Speech recognition

4] Computer Vision   **26 - 30**
    Definition of computer vision
    Image and video analysis
    Object detection and recognition

5] Robotics   **31 - 41**
    Definition of robotics
    Industrial robots
    Service robots
    Autonomous vehicles

6] Trending AI of 2023   42 - 61
    Tesla
    Neuralink
    Open AI
    Chat GPT
    Image Restorations AI
    Dalle AI
    Midjourney
    Stable Diffusion

7] Programming Languages of AI   62 - 84
    Python
    R
    Java
    C++
    Lisp
    Prolog

8] Applications of AI   85 - 98
    Healthcare
    Finance
    Transportation
    Customer service

9] Ethics and Society   99 - 110
    Bias in AI
    Job displacement
    Potential for misuse
    Regulation and governance

10] Future of AI   111 - 119
    Potential advancements
    Challenges and limitations
    Implications for society

11] Conclusion 120 - 126
    Summary of key points
    Future directions for AI research and development

//Introduction to Artificial Intelligence

Artificial intelligence, or AI, is the ability of a computer or machine to mimic or simulate human intelligence and behaviors. This includes the ability to learn, solve problems, make decisions, and communicate in natural language. The concept of AI has been around for decades, but recent advancements in technology have made it possible for AI systems to perform tasks that were once thought to be uniquely human.

There are different types of AI, including narrow or weak AI, which is designed for a specific task, and general or strong AI, which is designed to be able to perform any intellectual task that a human can. Most AI systems today fall into the category of narrow AI, as they are designed to perform a specific task, such as language translation or image recognition. However, the ultimate goal of AI research is to create systems that exhibit general intelligence and can adapt to new situations and tasks.

AI is achieved through the use of machine learning algorithms, which allow a machine to improve its performance on a task over time by learning from data and past experiences. Machine learning can be supervised, where the machine is provided with labeled data and a set of rules to follow, or unsupervised, where the machine is given a set of data and must find patterns and relationships within it on its own.

AI has the potential to revolutionize many industries and make tasks more efficient and accurate. It is already being used in a variety of fields, such as healthcare, finance, transportation, and customer service. However, the development and use of AI also raises ethical and societal concerns, including issues of bias, job displacement, and the potential for misuse.

{ Definition of AI

Artificial intelligence, or AI, is the field of computer science and engineering that aims to create intelligent machines that can think and act like humans. This includes the ability to learn, understand language, recognize patterns, make decisions, and solve problems. AI systems can be designed to perform a wide range of tasks, from simple to complex, and can adapt to new situations and changing environments.

AI is achieved through the use of algorithms that allow a machine to process and analyze data, and to improve its performance on a task over time through learning and experience. There are different types of AI, ranging from simple rule-based systems to complex machine learning algorithms that can make predictions and decisions without explicit instructions.

AI has the potential to transform many industries and improve our daily lives. It is already being used in fields such as healthcare, finance, transportation, and customer service. However, the development and use of AI also raises important ethical and societal questions, including issues of bias, job displacement, and the potential for misuse.

{ History of AI

The history of artificial intelligence (AI) dates back to ancient civilizations, when people attempted to create automatons and machines that could perform tasks and mimic human behaviors. However, the modern field of AI was officially founded in 1956 at a conference at Dartmouth College, where researchers gathered to discuss the possibility of creating machines that could think and learn.

In the decades that followed, AI research and development made significant progress, leading to the creation of expert systems, which were designed to perform tasks that required specialized knowledge, such as diagnosing medical conditions or solving complex mathematical problems. The field of machine learning, which allows machines to improve their performance on a task over time by learning from data and past experiences, also emerged during this time.

In the 1990s and 2000s, advances in computing power and the availability of large amounts of data made it possible for AI systems to perform tasks that were previously thought to be uniquely human, such as recognizing faces and translating languages. The development of the internet and the rise of big data also paved the way for the growth of AI in fields such as e-commerce and search engines.

Today, AI is being used in a variety of industries and applications, including healthcare, finance, transportation, and customer service. While the field of AI has come a long way, there are still many challenges and limitations to be addressed, and the ethical and societal implications of AI continue to be a subject of debate.

{ Types of AI

Artificial intelligence (AI) It can be classified into different types based on its capabilities and the tasks it is designed to perform. Here are some of the main types of AI:

1] Narrow or weak AI:
This type of AI is designed to perform a specific task or a set of related tasks. It is limited in scope and cannot adapt to new situations or perform tasks outside of its capabilities. Examples of narrow AI include virtual assistants, such as Siri and Alexa, which are designed to answer questions and perform tasks based on pre-programmed rules and information.

2] General or strong AI:
This type of AI is designed to exhibit general intelligence and the ability to perform any intellectual task that a human can. It is able to adapt to new situations and learn from experience, similar to human intelligence. However, creating a system that exhibits strong AI is a challenging task and has not yet been achieved.

3] Supervised learning:
This type of machine learning involves providing a machine with labeled data and a set of rules to follow in order to learn a task. The machine is given examples of correct inputs and outputs

and uses this information to make predictions or decisions. Supervised learning is useful for tasks such as image and speech recognition, where the machine can learn to identify patterns and features in the data.

4] Unsupervised learning:

This type of machine learning involves providing a machine with a set of data and allowing it to find patterns and relationships within the data on its own. The machine is not given any specific rules or examples of correct outputs, and must learn to identify patterns and trends in the data through exploration and analysis. Unsupervised learning is useful for tasks such as clustering and anomaly detection.

5] Reinforcement learning:

This type of machine learning involves training a machine to perform a task by providing it with rewards or punishments for certain actions. The machine learns to maximize its rewards by taking actions that lead to the desired outcome. Reinforcement learning is used in tasks such as game playing and autonomous control systems.

Overall, the different types of AI serve different purposes and have varying capabilities. Understanding the capabilities and limitations of each type is important for developing and using AI systems effectively.

//Machine Learning

Machine learning involves feeding a machine a large dataset and allowing it to learn from the data by identifying patterns and relationships within it. The machine can then use this learning to make predictions or decisions about new data, without being explicitly told how to do so. This allows the machine to adapt and improve its performance on the task over time, without the need for additional programming.

There are different types of machine learning, including supervised learning, unsupervised learning, and reinforcement learning. Supervised learning involves providing a machine with labeled data and a set of rules to follow, while unsupervised learning involves allowing the machine to find patterns and relationships within a dataset on its own. Reinforcement learning involves training a machine to perform a task by providing it with rewards or punishments for certain actions.

Machine learning has a wide range of applications, including image and speech recognition, natural language processing, and predictive analytics. It is being used in a variety of industries, including healthcare, finance, and e-commerce, and has the potential to transform many aspects of our lives. However, the development and use of machine learning also raises important ethical and societal questions, including issues of bias and the potential for misuse.

{ Definition of Machine Learning

Machine learning is a field of artificial intelligence (AI) that involves the use of algorithms and statistical models to allow a machine to improve its performance on a task over time by learning from data and past experiences. It is a type of AI that enables machines to learn and adapt automatically, without the need for explicit programming.

In machine learning, a machine is fed a large dataset and uses this data to learn about a specific task by identifying patterns and relationships within it. The machine can then use this learning to make predictions or decisions about new data, without being explicitly told how to do so. This allows the machine to adapt and improve its performance on the task over time, without the need for additional programming.

There are different types of machine learning, including supervised learning, unsupervised learning, and reinforcement learning. Each type involves a different approach to training the machine and can be used for different types of tasks and applications. Machine learning has a wide range of applications, including image and speech recognition, natural language processing, and predictive analytics, and is being used in a variety of industries.

{ Supervised Learning

Supervised learning is a type of machine learning in which a machine is provided with labeled training data and a set of rules to follow in order to learn a task. The machine is given examples of correct inputs and their corresponding outputs, and uses this information to make predictions or decisions about new data.

In supervised learning, the machine is "supervised" by the training data and the rules provided to it. It is able to learn from the data by finding patterns and relationships within it, and can use this learning to make predictions or decisions about new data.

Supervised learning is useful for tasks where there is a clear relationship between the input data and the output, and where the desired output is already known. Examples of tasks that can be solved using supervised learning include image and speech recognition, spam filtering, and predictive analytics.

Supervised learning algorithms can be further divided into two main categories: classification algorithms and regression algorithms. Classification algorithms are used to predict a categorical output, such as whether an email is spam or not, while regression algorithms are used to predict a continuous output, such as the price of a house based on its characteristics.

{ Unsupervised Learning

Unsupervised learning is a type of machine learning in which a machine is provided with a dataset and must find patterns and relationships within the data on its own, without being given specific rules or examples of correct outputs. The machine is not "supervised" in the same way as in supervised learning, and must explore the data and find structure within it on its own.

Unsupervised learning is useful for tasks where the desired output is not known in advance and there is no clear relationship between the input data and the output. Examples of tasks that can be solved using unsupervised learning include anomaly detection, clustering, and density estimation.

There are several types of algorithms that can be used for unsupervised learning, including clustering algorithms, which group data points into clusters based on their similarity, and dimensionality reduction algorithms, which reduce the number of features in a dataset while preserving as much information as possible.

Unsupervised learning can be useful for tasks such as data compression, anomaly detection, and data exploration, as it allows the machine to find patterns and relationships within the data that may not be obvious to a human observer. However, unsupervised learning can be more challenging than supervised learning, as the machine does not have the benefit of labeled examples to guide its learning.

{ Reinforcement Learning

Reinforcement learning is a type of machine learning in which a machine is trained to perform a task by receiving rewards or punishments for certain actions. The machine learns to take actions that maximize the rewards and minimize the punishments in order to achieve the desired outcome.

In reinforcement learning, the machine is referred to as an agent and the environment in which it operates is known as the environment. The agent takes actions within the environment and receives feedback in the form of rewards or punishments. The goal of the agent is to learn a policy, which is a set of rules that determine the actions it should take in order to maximize the rewards and achieve the desired outcome.

Reinforcement learning algorithms use trial and error to learn the best policy for a given task. The agent explores the environment and takes different actions, and the rewards or punishments it receives serve as feedback that helps it to learn the best actions to take in different situations.

Reinforcement learning has a wide range of applications, including game playing, autonomous control systems, and recommendation systems. It is a type of machine learning that can be used for tasks where the desired outcome is not always clear and the agent must learn through trial and error.

//Natural Language Processing

Natural Language Processing (NLP) is used for a wide range of tasks, including language translation, text classification, sentiment analysis, and language generation. It is an interdisciplinary field that combines linguistics, computer science, and artificial intelligence, and has applications in areas such as healthcare, customer service, and social media.

NLP algorithms are used to analyze and understand text and speech data in a way that is similar to how humans process language. They can recognize patterns, identify meaning, and generate responses in a way that is similar to human communication.

Some common techniques used in NLP include tokenization, which involves breaking down text into individual words or phrases; part-of-speech tagging, which involves identifying the role of each word in a sentence; and named entity recognition, which involves identifying and classifying named entities in text, such as people, organizations, and locations.

Overall, NLP is a rapidly growing field with a wide range of applications and the potential to transform the way humans and computers interact and communicate.

{ Definition Natural Language Processing

Natural language processing (NLP) is a field of artificial intelligence (AI) that involves enabling computers to understand, interpret, and generate human language. It involves the use of advanced algorithms and statistical models to analyze and understand the structure, meaning, and context of human language, and to generate responses in a way that is similar to human communication.

NLP is an interdisciplinary field that combines linguistics, computer science, and artificial intelligence, and has a wide range of applications in areas such as customer service, healthcare, and social media. It is used for tasks such as language translation, text classification, sentiment analysis, and language generation, and has the potential to revolutionize the way humans and computers interact and communicate.

NLP algorithms are designed to analyze and understand text and speech data in a way that is similar to how humans process language. They are able to recognize patterns, identify meaning, and generate responses in a way that is similar to human communication. Common techniques used in NLP include tokenization, part-of-speech tagging, and named entity recognition.

Overall, NLP is a rapidly growing field with a wide range of applications and the potential to transform the way humans and computers interact and communicate.

{ Text Classification

Text classification is a task in natural language processing (NLP) in which text data is automatically assigned to one or more predefined categories or labels. It involves the use of algorithms and statistical models to analyze and understand the content of a piece of text, and to determine its category or label based on certain features or characteristics.

Text classification is used for a wide range of applications, including spam filtering, sentiment analysis, and topic categorization. It is a common task in NLP and is useful for organizing and analyzing large volumes of text data.

There are several methods that can be used for text classification, including rule-based systems, which use a set of predefined rules to classify text; decision tree-based systems, which use a tree-like model to make decisions about the class of a piece of text; and machine learning-based systems, which use algorithms such as support vector machines and naive Bayes classifiers to learn from labeled training data and classify text based on its characteristics.

Overall, text classification is a useful tool for organizing and analyzing text data, and has a wide range of applications in areas such as customer service, healthcare, and social media.

Here are some examples of text classification:

1] Spam filtering:

Text classification can be used to identify and filter out spam emails based on certain characteristics, such as the presence of certain words or phrases, the use of all capital letters, or the use of foreign characters.

2] Sentiment analysis:

Text classification can be used to determine the sentiment or emotion expressed in a piece of text, such as positive, negative, or neutral. This can be useful for analyzing customer feedback or social media posts.

3] Topic categorization:

Text classification can be used to assign text data to predefined categories or topics, such as news articles, sports, or politics. This can be useful for organizing and searching large collections of text data.

4] Language translation:

Text classification can be used to identify the language of a piece of text, which can be useful for machine translation systems.

5] Fraud detection:

Text classification can be used to identify fraudulent transactions or activities based on certain characteristics, such as the presence of unusual patterns or language.

Overall, text classification has a wide range of applications and is a useful tool for organizing and analyzing text data.

{ Language Translation

Language translation is the process of converting text or speech from one language to another. It involves the use of natural language processing (NLP) algorithms and machine learning techniques to analyze and understand the structure, meaning, and context of a piece of text or speech in one language, and to generate a translation in another language that conveys the same meaning.

There are two main types of language translation: machine translation and human translation. Machine translation involves the use of software to automatically translate text or speech from one language to another, while human translation involves a person translating the text or speech manually.

Machine translation is becoming increasingly accurate and is widely used for tasks such as translating websites, documents, and social media posts. It is particularly useful for tasks that require fast turnaround times or involve large volumes of text. However, machine translation is not always perfect and may produce translations that are less accurate or natural-sounding than those produced by a human translator.

Human translation is generally more accurate and produces translations that are more natural-sounding than machine translation. It is typically used for tasks that require high levels of accuracy, such as legal documents and marketing materials, or for tasks that involve languages or subject matter that are more complex or nuanced.

Language translation is a useful tool for making information and content available to a wider audience and for facilitating communication between people who speak different languages.

Some examples of language translation:

1] Translating websites:

Language translation can be used to make websites and online content available to a wider audience by automatically translating it into multiple languages. This can be particularly useful for e-commerce websites, as it allows them to reach a global market.

2] Translating documents:

Language translation can be used to translate documents such as contracts, legal documents, and technical manuals, making them accessible to people who speak different languages.

3] Translating social media posts:

Language translation can be used to translate social media posts, comments, and reviews, making it easier for people who speak different languages to communicate and engage with each other online.

4] Translating speech:

Language translation can be used to translate spoken language in real-time, using technologies such as speech recognition and machine learning. This can be useful for tasks such as conference interpreting or for facilitating communication between people who speak different languages.

5] Translating subtitles:

Language translation can be used to translate subtitles for movies and TV shows, making them accessible to a wider audience.

Language translation is a useful tool for making information and content available to a wider audience and for facilitating communication between people who speak different languages.

{ Speech Recognition

Speech recognition is a technology that allows computers to understand and interpret human speech. It involves the use of artificial intelligence (AI) algorithms and machine learning techniques to analyze and understand the structure, meaning, and context of spoken language, and to generate a written or typed translation of the speech.

Speech recognition is becoming increasingly accurate and is used for a wide range of applications, including voice-controlled assistants, language translation, and transcription. It is particularly useful for tasks that involve large volumes of spoken data or for situations where it is more convenient or efficient to speak rather than type.

There are two main types of speech recognition: isolated word recognition and continuous speech recognition. Isolated word recognition involves the recognition of individual words that are spoken one at a time, while continuous speech recognition involves the recognition of continuous, uninterrupted speech. Continuous speech recognition is generally more challenging, as it requires the system to recognize the words and their context within a larger stream of speech.

Speech recognition is a rapidly growing field with a wide range of applications and the potential to transform the way humans and computers communicate and interact.

There are a number of ways that you can interact with speech recognition technology. Here are a few examples:

1] Voice-controlled assistants:

Many devices, such as smartphones and smart speakers, have built-in voice-controlled assistants that allow you to interact with them using spoken commands. You can use these assistants to perform tasks such as making phone calls, setting reminders, and playing music.

2] Language translation:

Some language translation tools use speech recognition to allow you to speak in one language and have your words translated and spoken in another language. This can be useful for tasks such as interpreting or for facilitating communication between people who speak different languages.

3] Transcription:

There are a number of tools that use speech recognition to transcribe spoken words into written or typed text. These tools can be useful for tasks such as transcribing meetings or lectures, or for creating written records of spoken conversations.

4] Speech-to-text software:

There are a number of software programs that use speech recognition to convert spoken words into written or typed text. These programs can be useful for tasks such as dictation or for creating written documents from spoken notes.

Overall, there are many ways to interact with speech recognition technology and it has a wide range of applications in areas such as language translation, customer service, and transcription.

Speech recognition is a rapidly growing field with many potential uses and applications, and it is likely to have an increasingly significant impact on a wide range of industries in the coming years.

There are many speech recognition systems available today, ranging from simple voice-controlled assistants to complex transcription and translation tools. Here are a few examples of existing speech recognition systems:

1] Voice-controlled assistants:

Many devices, such as smartphones and smart speakers, have built-in voice-controlled assistants that allow you to interact with them using spoken commands. Examples of these assistants include **Apple's Siri, Amazon's Alexa, and Google's Assistant**.

2] Language translation:

There are a number of language translation tools that use speech recognition to allow you to speak in one language and have your words translated and spoken in another language. Examples of these tools include **Google Translate and Microsoft Translator**.

3] Transcription:

There are a number of tools that use speech recognition to transcribe spoken words into written or typed text. Examples of these tools include **Otter.ai and Rev.com.**

4] Speech-to-text software:

There are a number of software programs that use speech recognition to convert spoken words into written or typed text. Examples of these programs include **Dragon NaturallySpeaking and Windows Speech Recognition.**

//Computer Vision

Computer vision is used for a wide range of applications, including image and video analysis, object and facial recognition, and autonomous systems. It is an interdisciplinary field that combines computer science, machine learning, and engineering, and has applications in areas such as healthcare, transportation, and manufacturing.

Computer vision algorithms are used to analyze and understand visual data in a way that is similar to how humans process and interpret visual information. They are able to recognize patterns, identify objects and features, and extract meaning from images and video.

Some common techniques used in computer vision include feature extraction, which involves identifying and extracting important features or characteristics from images; image segmentation, which involves dividing an image into distinct regions or segments; and object recognition, which involves identifying and classifying objects in an image.

{ Definition of Computer Vision

Computer vision is a field of artificial intelligence (AI) that deals with the design and development of algorithms and systems that can recognize and interpret visual data. It involves the use of machine learning and computer vision techniques to analyze and understand images and video, and to extract meaningful information from them.

Computer vision is also defined as a field of artificial intelligence that focuses on enabling computers to interpret and understand visual data, such as images and video. It involves the development of algorithms and systems that can analyze and understand visual information in a way that is similar to how humans process and interpret visual information.

Using techniques such as machine learning and feature extraction, computer vision algorithms are able to identify and classify objects, recognize patterns, and extract meaning from images and video. This technology has a wide range of applications, including object and facial recognition, image and video analysis, and the development of autonomous systems.

Computer vision is an interdisciplinary field that combines computer science, machine learning, and engineering, and has the potential to revolutionize the way computers and humans interact with the world around them.

{ Image and Video Analysis

Image and video analysis is the process of using algorithms and techniques to extract meaning and information from images and video. It is a subfield of computer vision, which is a field of artificial intelligence (AI) that deals with the design and development of algorithms and systems that can recognize and interpret visual data.

Image and video analysis involves the use of machine learning and computer vision techniques to analyze and understand visual data, and to extract meaningful information from it. It is used for a wide range of applications, including object and facial recognition, image and video classification, and the development of autonomous systems.

There are many techniques that can be used for image and video analysis, including feature extraction, which involves identifying and extracting important features or characteristics from images; image segmentation, which involves dividing an image into distinct regions or segments; and object recognition, which involves identifying and classifying objects in an image.

Image and video analysis is the process of using algorithms and techniques to extract meaning and information from images and video. It has a wide range of applications in areas such as healthcare, transportation, and manufacturing. Here are a few examples of how image and video analysis is used in real life:

1] Facial recognition:
Image and video analysis can be used for facial recognition, which involves identifying and verifying the identity of individuals from images or video. This technology is used in a variety of applications, including security and surveillance, social media, and customer service.

2] Object recognition:

Image and video analysis can be used to recognize and classify objects in images or video. This technology is used in applications such as self-driving cars, which use image and video analysis to identify and classify objects on the road.

3] Medical imaging:

Image and video analysis is used in medical imaging to analyze and interpret medical images such as X-rays, CT scans, and MRI scans. This technology can be used to identify and diagnose diseases, assess the effectiveness of treatments, and guide surgical procedures.

4] Video surveillance:

Image and video analysis is used in video surveillance systems to automatically identify and track individuals and objects in real-time. This technology is used in applications such as security and public safety, and can be used to detect and prevent crime or to identify potential threats.

Actual Applications in Real Life:

Government agencies: Image and video analysis is used by government agencies for a variety of purposes, including security and surveillance, public safety, and intelligence gathering.

Healthcare organizations: Image and video analysis is used by healthcare organizations for tasks such as medical imaging, which involves analyzing and interpreting medical images to identify and diagnose diseases, assess the effectiveness of treatments, and guide surgical procedures.

Transportation companies: Image and video analysis is used by transportation companies to develop autonomous systems, such as self-driving cars, which use image and video analysis to identify and classify objects on the road.

Manufacturing companies: Image and video analysis is used by manufacturing companies to improve efficiency and productivity by automating tasks such as quality control and inspection.

Retail companies: Image and video analysis is used by retail companies to improve customer experiences, such as by using facial recognition to personalize advertisements or to optimize product displays.

Image and video analysis is a widely used technology that has a wide range of applications and is used by a variety of organizations and industries.

//Robotics

Robotics is the Future of our human kind. It is a rapidly growing field with a wide range of applications and the potential to transform the way humans and computers interact with the world around them. It is an interdisciplinary field that combines computer science, engineering, and machine learning, and has the potential to revolutionize industries and change the way we live and work.

There are many types of robots, including humanoid robots, which are designed to resemble and mimic human behavior; industrial robots, which are used in manufacturing and assembly lines; and mobile robots, which are able to move and navigate in the physical world.

{ Definition of Robotics

Robotics is the branch of artificial intelligence (AI) that deals with the design and development of robots, which are automated systems that can perform tasks independently or in collaboration with humans. It involves the use of machine learning and AI algorithms to enable robots to perceive, reason, and act in the physical world.

Robots are used in a wide range of applications, including manufacturing, healthcare, transportation, and military. They can be programmed to perform a variety of tasks, such as assembly, inspection, transportation, and surgery, and can be designed to operate in a variety of environments, including land, sea, air, and space.

Robotics is a rapidly growing field with a wide range of applications and the potential to transform the way humans and computers interact with the world around them. It is an interdisciplinary field that combines computer science, engineering, and machine learning, and has the potential to revolutionize industries and change the way we live and work.

{ Industrial Robots

Industrial robots are automated systems that are used in manufacturing and assembly lines to perform tasks such as welding, painting, and assembly. They are programmed to perform specific tasks and can operate 24 hours a day, seven days a week, without the need for breaks or time off.

Industrial robots are typically designed to operate in a factory or other industrial setting and are used to improve efficiency, productivity, and quality. They can be programmed to perform a wide range of tasks and can be customized to meet the specific needs of different industries.

There are several types of industrial robots, including:

Cartesian robots: These robots are characterized by their rectangular workspace and are typically used for tasks such as welding and painting.

Cylindrical robots: These robots are characterized by their cylindrical workspace and are typically used for tasks such as assembly and handling.

Spherical robots: These robots are characterized by their spherical workspace and are typically used for tasks such as handling and inspection.

Industrial robots are an important part of the manufacturing industry and are used to improve efficiency, productivity, and quality. They are a rapidly growing field with a wide range of applications and the potential to transform the way humans and computers interact with the world around them.

The future of robotics is an area of active research and development, and there are many potential developments and applications that are currently being explored. Here are a few examples of what the future of robotics might look like:

Increased automation: Robotics technology is likely to continue to advance and become more widespread, leading to increased automation in a variety of industries. This could result in significant changes to the way work is done, including the displacement of some jobs by robots.

Increased collaboration between humans and robots: As robots become more advanced and sophisticated, it is likely that they will be able to work more closely with humans, enabling more efficient and effective collaboration. This could lead to the development of new roles and job opportunities that involve working with robots.

Increased use of robots in healthcare: Robots are likely to play an increasingly important role in healthcare in the future, performing tasks such as assisting with surgery and delivering medication to patients.

Increased use of robots in transportation: As autonomous vehicles become more widespread, robots are likely to play an increasingly important role in transportation, potentially revolutionizing the way we travel.

Increased use of robots in hazardous environments: Robots are likely to be used more widely in hazardous environments, such as in disaster response or in exploration of extreme environments, where it is too dangerous for humans to go.

{ Service Robots

Service robots are automated systems that are designed to perform tasks that assist or benefit humans in some way. They are used in a variety of settings, including homes, hospitals, schools, and offices.

There are many types of service robots, including:

Domestic robots: These robots are designed to perform tasks in the home, such as cleaning, cooking, and assisting with household chores.

Educational robots: These robots are designed to be used in educational settings, such as schools, to assist with teaching and learning.

Medical robots: These robots are designed to be used in healthcare settings, such as hospitals, to assist with tasks such as medication delivery and surgery.

Professional service robots: These robots are designed to be used in professional settings, such as offices, to assist with tasks such as customer service and data entry.

Service robots are an increasingly important part of our daily lives, and they have the potential to improve efficiency, productivity, and quality of life in a variety of settings. They are a rapidly growing field with a wide range of applications and the potential to transform the way humans and computers interact with the world around them.

Service robots are related to artificial intelligence (AI) in that they are automated systems that are designed to perform tasks that assist or benefit humans in some way. They use AI algorithms and techniques to enable them to perceive, reason, and act in the physical world, and to make decisions and take actions based on that information.

Service robots are typically designed to operate independently or in collaboration with humans, and they can be programmed to perform a wide range of tasks. They use AI algorithms and techniques such as machine learning, natural language processing, and computer vision to enable them to understand and respond to their environment and to the needs of humans.

Service robots are an important part of the field of AI, and they have the potential to transform the way humans and computers interact with the world around them. They are a rapidly growing field with a wide range of applications and the potential to revolutionize industries and change the way we live and work.

{ Autonomous Vehicles

Autonomous vehicles, also known as self-driving vehicles, are vehicles that are capable of navigating and operating without human input. They use a variety of technologies, including artificial intelligence (AI), machine learning, and sensors, to enable them to perceive and understand their environment and to make decisions about how to navigate and operate.

There are several levels of autonomy that autonomous vehicles can achieve, ranging from fully autonomous (level 5) to partially autonomous (levels 1-4). Fully autonomous vehicles are capable of operating without any human intervention, while partially autonomous vehicles require some level of human input or supervision.

Autonomous vehicles are being developed for a variety of applications, including personal transportation, public transportation, and commercial transportation. They have the potential to revolutionize the way we travel and to improve safety, efficiency, and accessibility.

However, there are also challenges and concerns associated with autonomous vehicles, including issues related to safety, privacy, and liability. As a result, the development and deployment of autonomous vehicles is an active area of research and development, and there are many regulatory and ethical issues that need to be considered.

Overall, autonomous vehicles are a rapidly growing field with a wide range of potential applications and the potential to transform the way humans and computers interact with the world around them.

There are already a number of autonomous vehicles on the road today, and the number is likely to continue to grow in the coming years. Autonomous vehicles, also known as self-driving vehicles, are vehicles that are capable of navigating and operating without human input. They use a variety of technologies, including artificial intelligence (AI), machine learning, and sensors, to enable them to perceive and understand their environment and to make decisions about how to navigate and operate.

There are several levels of autonomy that autonomous vehicles can achieve, ranging from fully autonomous (level 5) to partially autonomous (levels 1-4). Fully autonomous vehicles are capable of operating without any human intervention, while partially autonomous vehicles require some level of human input or supervision.

There are already a number of autonomous vehicle projects and pilot programs being conducted around the world, and some companies are planning to launch fully autonomous vehicles in the near future. However, there are also challenges and concerns associated with autonomous vehicles, including issues related to safety, privacy, and liability. As a result, the development and deployment of autonomous vehicles is an active area of research and development, and there are many regulatory and ethical issues that need to be considered.

There are several companies that are working on developing autonomous vehicles. Here are a few examples:

Tata Motors: Tata Motors is an Indian multinational automotive company that is working on developing autonomous vehicles. The company has partnerships with several technology companies, including Microsoft and NVIDIA, to develop autonomous driving technology.

Mahindra & Mahindra: Mahindra & Mahindra: is an Indian multinational conglomerate that is involved in a variety of industries, including automotive. The company is working on developing autonomous vehicles for a variety of applications, including personal transportation, public transportation, and commercial transportation.

Ashok Leyland: Ashok Leyland: is an Indian automotive company that is a subsidiary of the Hinduja Group. The company is working on developing autonomous vehicles for a variety of applications, including personal transportation, public transportation, and commercial transportation.

Tech Mahindra: Tech Mahindra is an Indian multinational technology company that is working on developing autonomous vehicles for a variety of applications, including personal transportation and public transportation.

Waymo: Waymo is a self-driving technology company that is a subsidiary of Alphabet, the parent company of Google. Waymo is developing autonomous vehicles for a variety of applications, including personal transportation, public transportation, and commercial transportation.

Tesla: Tesla is a company that is known for its electric vehicles, and it has been working on developing autonomous driving technology for its cars. Tesla's Autopilot feature is a partially autonomous driving system that is available on some of its vehicles.

Cruise: Cruise is a self-driving technology company that is a subsidiary of General Motors. Cruise is developing autonomous vehicles for a variety of applications, including personal transportation and commercial transportation.

Navya: Navya is a company that is developing autonomous electric shuttles for use in public transportation. The company's autonomous shuttles have already been deployed in several cities around the world.

EZ10: EZ10 is an autonomous electric shuttle developed by the French company EasyMile. The EZ10 has been deployed in a number of cities around the world for use in public transportation.

The future of autonomous vehicles is an active area of research and development, and there are many potential developments and applications that are currently being explored. Here are a few examples of what the future of autonomous vehicles might look like:

Increased use of autonomous vehicles: As autonomous vehicle technology continues to advance, it is likely that autonomous vehicles will become more widespread and will be used in a variety of applications, including personal transportation, public transportation, and commercial transportation.

Increased safety: Autonomous vehicles are likely to improve safety on the roads, as they are less likely to make mistakes or to be distracted than human drivers. This could lead to a significant reduction in traffic accidents and fatalities.

Increased efficiency: Autonomous vehicles are likely to improve efficiency by enabling the optimization of routes, reducing traffic congestion, and enabling more efficient use of road space.

Increased accessibility: Autonomous vehicles are likely to improve accessibility for people who are unable to drive, such as the elderly or people with disabilities.

Increased environmental sustainability: Autonomous vehicles are likely to be more environmentally friendly, as they are likely to be electric and will be able to optimize routes and speeds to reduce energy consumption.

Overall, the future of autonomous vehicles is an exciting and rapidly growing field with many potential developments and applications. It has the potential to transform the way we travel and to improve safety, efficiency, and accessibility. However, there are also challenges and concerns associated with autonomous vehicles, including issues related to safety, privacy, and liability, that will need to be addressed.

//Trending AI of 2023

As we approach **2023**, there are a number of trends in the **artificial intelligence (AI)** field that are worth paying attention to. These trends are shaping the future of AI and will likely have a significant impact on a wide range of industries and applications.

One trend to watch is the continued development of **artificial general intelligence (AGI)**. AGI refers to the development of AI systems that are able to perform a wide range of tasks at the same level of human intelligence. While we are still some way off from achieving true AGI, significant progress has been made in recent years and it is likely that we will see continued advancements in this area over the next few years.

Another trend to keep an eye on is the growth of **natural language processing (NLP)**. NLP is the ability of computers to understand, interpret, and generate human language. This is a key area of AI research, as it has a wide range of potential applications, including the development of more sophisticated and human-like chatbots and virtual assistants.

Machine learning (ML) is also expected to continue to be a major trend in the AI field. ML involves the use of algorithms and statistical models to enable computers to learn from data without being explicitly programmed. ML has already been applied to a wide range of tasks, including prediction, classification, and recommendation, and it is likely that we will see further advancements in this area over the next few years.

Robotics is another area of AI that is likely to see significant growth in the coming years. The integration of AI and machine learning in robotics is enabling the development of more advanced systems that are able to recognize and manipulate objects, as well as navigate and interact with their environment.

Finally, the growth of autonomous vehicles is another trend to watch in the AI field. Autonomous vehicles use a combination of sensors, cameras, and AI to enable self-driving capabilities, and they are expected to become more widespread in the coming years.

These are just a few of the trends that we can expect to see in the AI field over the next few years. It is an exciting time for AI research and development, and we can look forward to seeing a range of exciting new technologies and applications emerge in the coming years.

{ Tesla

Tesla is a company that has revolutionized the automotive industry with its innovative electric vehicles (EVs) and **advanced driver assistance systems (ADAS)**. Founded in 2003 by CEO Elon Musk, Tesla has quickly become a leader in the production of electric vehicles, with a mission to accelerate the world's transition to sustainable energy.

One of the standout features of Tesla's vehicles is their advanced driver assistance systems, which use a combination of sensors, cameras, and AI to enable semi-autonomous driving capabilities.

Tesla's Autopilot system, for example, uses radar, ultrasonic sensors, and cameras to detect and track the surrounding environment, allowing the vehicle to automatically steer, accelerate, and brake in response to traffic and road conditions.

In addition to Autopilot, Tesla offers a number of other advanced driver assistance features, including Navigate on Autopilot, which uses real-time traffic data to automatically plan the most efficient route to a destination, and Summon, which allows the driver to remotely move the vehicle in and out of tight parking spaces using their smartphone.

Another key feature of Tesla's vehicles is their electric powertrain, which uses advanced lithium-ion battery technology to provide long range and fast charging capabilities. Tesla's vehicles are designed to be energy efficient, with a focus on maximizing the distance that can be traveled on a single charge. The company's Supercharger network, which consists of over 20,000 charging stations worldwide, allows Tesla drivers to quickly and easily charge their vehicles on the go.

In addition to their advanced technology and sustainability features, Tesla's vehicles are also known for their sleek and stylish design. The company's lineup includes a range of vehicles to suit different needs and preferences, including the Model S, a luxury sedan; the Model X, a full-size SUV; and the Model 3, a more affordable midsize sedan.

Overall, Tesla's vehicles represent the future of the automotive industry, with their advanced technology, sustainability, and style.

As the company continues to innovate and expand its product line, it is likely that Tesla will remain a leader in the EV market for years to come.

Future Plans Of Tesla are:

Tesla, Inc. is an American multinational corporation that specializes in electric vehicles, energy storage and solar panel manufacturing based in Palo Alto, California. Founded in 2003, the company specializes in electric cars, lithium-ion battery energy storage, and residential photovoltaic panels (through the subsidiary company named Tesla Energy) and home batteries. Tesla first gained widespread attention following its production of the Tesla Roadster, the first electric sports car, in 2008. The company's second vehicle, the Model S, an electric luxury sedan, debuted in 2012 and is built at the Tesla Factory in California. The Model S was followed by the Model X, a crossover SUV, in 2015, and the Model 3, a sedan designed for the mass market, in 2017. In March 2018, Tesla released a semi-autonomous vehicle, the Model S 75D.

In the **future**, Tesla plans to expand its product line to include electric semi trucks and pickup trucks. The company is also working on developing new technologies in the areas of AI and machine learning, with a focus on improving the performance and safety of its vehicles. In addition, Tesla is exploring new markets and partnerships, including collaborations with other automakers and energy companies. Finally, Tesla is continuing to invest in its manufacturing and supply chain infrastructure, with a goal of increasing production efficiency and reducing costs.

Overall, Tesla's future plans involve continued growth and expansion in the electric vehicle market, as well as the development of new technologies and partnerships to support its long-term goals.

{ Neuralink

Neuralink is a neurotechnology company that was founded in 2016 by Elon Musk and others with the goal of developing implantable **brain–machine interfaces (BMIs)**. The company aims to create devices that can be implanted in the human brain and connected to a computer, allowing people to directly communicate with machines and other electronic devices.

One of the main goals of Neuralink is to create a BMI that can be used to treat neurological conditions, such as Parkinson's disease, Alzheimer's disease, and spinal cord injuries. By providing a way for the brain to communicate directly with devices, such as prosthetics or exoskeletons, Neuralink's BMI could potentially restore lost function and improve the lives of people with these conditions.

In addition to its medical applications, Neuralink is also exploring the use of BMIs for more general purposes, such as enhancing human intelligence and abilities. For example, the company is working on developing a BMI that could be used to improve memory, learning, and other cognitive functions. Some experts believe that BMIs could eventually be used to enhance human intelligence to the point where people are able to perform complex tasks at superhuman speeds or even access new dimensions of thought and perception.

Neuralink's BMI is still in the early stages of development, and it is not yet clear when the technology will be ready for widespread use. However, the company has made significant progress in recent years, and it is likely that we will see continued advancements in this area over the next few years.

Overall, Neuralink's project represents an exciting and potentially revolutionary advancement in the field of neurotechnology. If successful, the company's BMI could have a significant impact on the way we think about and interact with the world around us.

Benefits of Neuralink for Humans are:
There are a number of potential benefits that Neuralink's brain–machine interfaces (BMIs) could offer to humans. Some of the main benefits include:

1] Medical treatment:
One of the primary goals of Neuralink is to use BMIs to treat neurological conditions, such as Parkinson's disease, Alzheimer's disease, and spinal cord injuries. By providing a way for the brain to communicate directly with devices, such as prosthetics or exoskeletons, Neuralink's BMI could potentially restore lost function and improve the lives of people with these conditions.

2] Enhanced cognition:
Neuralink is also exploring the use of BMIs to enhance human intelligence and abilities. For example, the company is

working on developing a BMI that could be used to improve memory, learning, and other cognitive functions.

3] Improved communication:

BMIs could also be used to improve communication for people with certain conditions, such as locked-in syndrome or severe aphasia. By providing a direct connection between the brain and a computer, BMIs could allow people to communicate their thoughts and intentions even if they are unable to speak or move.

4] Increased accessibility:

BMIs could also potentially be used to make technology more accessible for people with disabilities. For example, a BMI could allow people with mobility impairments to control devices or access information using their thoughts rather than physical actions.

Overall, Neuralink's BMIs could offer a range of benefits to humans, including the ability to treat neurological conditions, enhance cognitive abilities, improve communication, and increase accessibility.

{ Open AI

OpenAI is an **artificial intelligence (AI)** research laboratory consisting of the for-profit corporation OpenAI LP and its parent company, the non-profit OpenAI Inc. The company was founded in 2015 by a group of entrepreneurs and researchers, including

Elon Musk, Sam Altman, and Greg Brockman, with the goal of advancing the field of AI and promoting its responsible development.

One of the main areas of focus for OpenAI is the development of artificial general intelligence (AGI), which refers to the creation of AI systems that are able to perform a wide range of tasks at the same level of human intelligence. OpenAI researchers are working on a number of different approaches to AGI, including machine learning, neural networks, and evolutionary algorithms.

In addition to its AGI research, OpenAI is also working on a number of other projects related to AI. These include the development of AI systems for natural language processing, computer vision, and robotics, as well as research on the ethical and societal implications of AI.

One of the key goals of OpenAI is to make its research and technology available to the wider community. To this end, the company has released a number of open source tools and libraries, including the popular deep learning library PyTorch and the GPT-3 language model. These tools have been widely adopted by researchers and developers around the world, and have contributed to the advancement of the field of AI.

Overall, OpenAI is a leading research laboratory in the field of AI, and its work is shaping the future of the technology. With its focus on AGI and its commitment to open source research and technology, OpenAI is playing a key role in advancing the field of AI and helping to ensure its responsible development.

OpenAI is an **artificial intelligence (AI)** research laboratory that has developed a number of tools and libraries to advance the field of AI and make its research and technology available to the wider community.

Some of the main types of tools that have been developed by OpenAI include:

1] Deep learning libraries:

OpenAI has developed several deep learning libraries, including PyTorch, which is a widely used open source library for machine learning and deep learning.

2] Language models:

OpenAI has developed several language models, including the GPT-3 model, which is a large-scale language model that can generate human-like text.

3] Robotics frameworks:

OpenAI has developed several robotics frameworks, including the MuJoCo physics engine, which is a software toolkit for simulating and analyzing the dynamics of rigid body systems.

4] Computer vision tools:

OpenAI has developed a number of tools for computer vision, including the DALL-E image generation model, which can generate images from text descriptions.

5] Tools for ethical and societal implications of AI:

OpenAI has also developed a number of tools and resources related to the ethical and societal implications of AI, including the AI Safety research group, which focuses on the safety and alignment of AI with human values.

Overall, OpenAI has developed a range of tools and resources to advance the field of AI and make its research and technology available to the wider community. These tools and resources are being used by researchers and developers around the world to advance the field of AI and help ensure its responsible development.

{ Chat GPT

GPT-3 (short for "Generative Pre-trained Transformer 3") is a natural language processing (NLP) model developed by OpenAI. It is one of the largest and most powerful NLP models to date, with 175 billion parameters (weights or "knobs" that the model can adjust during training).

GPT-3 is a type of language model, which means that it is trained to predict the next word in a given sequence of words. It does this by analyzing a large dataset of text and learning to identify patterns and relationships between words and phrases. This allows GPT-3 to generate human-like text that is coherent and flows naturally.

One of the key features of GPT-3 is its ability to perform a wide range of NLP tasks without the need for explicit task-specific programming. This is known as "few-shot learning," and it allows

GPT-3 to adapt to new tasks quickly and effectively. For example, GPT-3 can be trained to perform tasks such as translation, summarization, and question answering by simply providing it with a few examples of what the task should look like.

GPT-3 has a number of potential applications, including:

1] Text generation:
GPT-3 can be used to generate human-like text for a wide range of applications, such as generating news articles, creating marketing content, or writing fiction.

2] Translation:
GPT-3 can be trained to perform machine translation, allowing it to translate text from one language to another.

3] Summarization:
GPT-3 can be used to generate summaries of long documents or articles, making it easier for people to quickly understand their main points.

4] Question answering:
GPT-3 can be used to answer questions posed in natural language, making it possible to build intelligent virtual assistants or chatbots.

5] Text classification:

GPT-3 can be used to classify text into different categories, such as spam or legitimate emails, or to identify the sentiment (positive, negative, or neutral) expressed in a piece of text.

Overall, GPT-3 is a powerful NLP model that has the potential to revolutionize the way we interact with and process language. Its ability to perform a wide range of NLP tasks without explicit task-specific programming makes it a valuable tool for a variety of applications, including text generation, translation, summarization, question answering, and text classification.

{ Image Restorations AI

Image restoration is the process of repairing or improving the quality of damaged or degraded images. This can involve removing noise, blur, or other distortions from the image, as well as filling in missing or damaged pixels. Image restoration is an important task in a variety of fields, including photography, medical imaging, and remote sensing.

In recent years, **artificial intelligence (AI)** has emerged as a powerful tool for image restoration. AI-based image restoration techniques are able to learn patterns and features from a large dataset of images and use this knowledge to restore damaged or degraded images.

There are a number of different approaches to image restoration using AI, including:

1] Deep learning:

Deep learning is a type of machine learning that uses neural networks to learn complex patterns and features from data. Deep learning has been applied to image restoration by training neural networks on large datasets of images and using the resulting model to restore damaged or degraded images.

2] Generative adversarial networks (GANs):

GANs are a type of neural network that consists of two networks: a generator network and a discriminator network. The generator network is trained to generate new images, while the discriminator network is trained to distinguish between real and generated images. GANs have been applied to image restoration by training the generator network to generate images that are similar to the damaged or degraded input image, while the discriminator network is trained to distinguish between the generated image and the original image.

3] Non-local means (NLM):

NLM is a type of image restoration technique that is based on the idea that similar patches of pixels in an image are likely to be corrupted by similar noise or blur. NLM uses this idea to restore images by replacing damaged or degraded pixels with similar pixels from other parts of the image. AI has been used to improve the performance of NLM by learning the patterns and features of the image and using this knowledge to guide the restoration process.

AI-based image restoration techniques have a number of advantages over traditional methods. For example, they are able to learn patterns and features from data, which allows them to adapt to different types of damage or degradation. They are also able to process images in real-time, making them suitable for use in applications such as video restoration or real-time image processing.

Overall, AI is a powerful tool for image restoration, offering the ability to learn patterns and features from data and adapt to different types of damage or degradation. It has a wide range of applications, including photography, medical imaging, and remote sensing, and has the potential to revolutionize the way we process and restore images.

{ Dalle AI

DALL-E (pronounced "dolly") is an artificial intelligence (AI) system developed by OpenAI. It is a deep learning model that is able to generate images from textual descriptions, such as "a two-story pink house with a white fence and a red door." DALL-E is trained on a dataset of text-image pairs and uses this training to generate images that match the descriptions provided to it.

One of the key features of DALL-E is its ability to generate a wide range of images, including ones that do not exist in the real world. For example, it can generate images of animals with multiple heads or objects with impossible shapes. This ability to generate novel and creative images sets DALL-E apart from other

AI image generation systems, which are typically limited to generating images of objects or scenes that exist in the real world.

DALL-E has a number of potential applications, including:

1] Image generation:

DALL-E can be used to generate images from text descriptions, making it possible to create images of objects or scenes that do not exist in the real world.

2] Artistic inspiration:

DALL-E can be used by artists as a source of inspiration, allowing them to generate a wide range of ideas for new works of art.

3] Product design:

DALL-E can be used by product designers to generate ideas for new products or to visualize how a product might look in different colors or configurations.

4] Advertising:

DALL-E can be used to generate images for use in advertising campaigns, allowing companies to quickly and easily create a wide range of visual content.

In addition to its potential applications, DALL-E has also attracted attention for its ability to generate images that are both novel and unsettling. Some of the images generated by DALL-E are so strange or surreal that they are difficult to interpret or understand.

This has led to debates about the potential impact of AI on creativity and art, and whether AI systems like DALL-E should be given more or less control over the creative process.

Overall, DALL-E is a unique and powerful AI system that has the ability to generate a wide range of images from text descriptions. Its ability to generate novel and creative images sets it apart from other AI image generation systems, and it has a number of potential applications in fields such as art, product design, and advertising. However, its ability to generate strange and unsettling images has also raised questions about the potential impact of AI on creativity and art, and the role that AI should play in the creative process.

{ Midjourney

Midjourney is an artificial intelligence (AI) platform developed by the startup company of the same name. It is designed to enable businesses to automate repetitive tasks and processes using AI and machine learning technologies.

One of the key features of Midjourney is its ability to learn and adapt to new tasks and processes over time. By training the platform on a dataset of tasks and processes, businesses can use Midjourney to automate a wide range of tasks, such as data entry, customer service, and marketing. As the platform learns and adapts to new tasks and processes, it becomes more efficient and effective at automating them, leading to cost and time savings for the business.

Midjourney is designed to be easy to use and requires minimal technical expertise to set up and use. It includes a range of tools and features that make it simple to train the platform on new tasks and processes, including a visual interface and a range of pre-built models and algorithms. This makes it easy for businesses to get started with Midjourney and begin automating tasks and processes quickly.

In addition to its ability to automate tasks and processes, Midjourney also includes a range of tools and features for managing and monitoring the performance of the platform. This includes dashboards and analytics tools that allow businesses to track the progress of tasks and processes, as well as identify areas for improvement.

Overall, Midjourney is a powerful AI platform that enables businesses to automate repetitive tasks and processes using machine learning and AI technologies. Its ability to learn and adapt to new tasks and processes over time makes it an effective tool for cost and time savings, and its easy-to-use interface and range of management and monitoring tools make it simple to set up and use.

Some potential uses of Midjourney include:

1] Data entry:
Midjourney can be used to automate the process of entering data into databases or other systems. By training the platform on a dataset of data entry tasks, businesses can use Midjourney to

quickly and accurately enter large amounts of data, saving time and reducing the risk of errors.

2] Customer service:

Midjourney can be used to automate customer service tasks, such as answering customer inquiries or handling complaints. By training the platform on a dataset of customer service interactions, businesses can use Midjourney to provide fast and accurate responses to customer inquiries, improving customer satisfaction.

3] Marketing:

Midjourney can be used to automate marketing tasks, such as creating and sending email campaigns or social media posts. By training the platform on a dataset of marketing tasks, businesses can use Midjourney to create and execute marketing campaigns more efficiently, saving time and resources.

4] Supply chain management:

Midjourney can be used to automate tasks and processes related to supply chain management, such as ordering inventory or tracking shipments. By training the platform on a dataset of supply chain management tasks, businesses can use Midjourney to streamline and optimize their supply chain operations.

5] Human resources:

Midjourney can be used to automate tasks and processes related to human resources, such as processing employee applications or scheduling interviews. By training the platform on a dataset of human resources tasks, businesses can use Midjourney to improve the efficiency of their HR operations.

Overall, Midjourney is a versatile AI platform that can be used to automate a wide range of tasks and processes in a variety of industries and contexts. Its ability to learn and adapt to new tasks and processes over time makes it an effective tool for cost and time savings, and its easy-to-use interface and range of management and monitoring tools make it simple to set up and use.

{ Stable Diffusion

Stable Diffusion is a machine learning algorithm developed by researchers at the Massachusetts Institute of Technology (MIT). It is designed to enable machine learning models to continuously learn and adapt to new data over time, without the need for retraining.

One of the key challenges in machine learning is the need to periodically retrain models on new data in order to maintain their accuracy and effectiveness. This can be time-consuming and resource-intensive, especially for large models or datasets. Stable Diffusion aims to address this issue by enabling models to continuously learn and adapt to new data in a stable and efficient manner.

The Stable Diffusion algorithm works by adding new data to the model in a controlled way, using a diffusion process that ensures that the model remains stable and accurate as it adapts to the new data. This allows the model to continuously learn and adapt to new data over time, without the need for retraining.

In addition to enabling continuous learning, Stable Diffusion also has a number of other benefits. It can significantly reduce the amount of data and computation required for training machine learning models, making it more efficient and cost-effective. It can also improve the generalization performance of machine learning models, enabling them to better handle new and unseen data.

Stable Diffusion has the potential to revolutionize machine learning and enable the development of more advanced and effective models. Its ability to enable continuous learning and adaptability in machine learning models has wide-ranging applications, including in industries such as healthcare, finance, and transportation.

Overall, Stable Diffusion is a promising machine learning algorithm that has the potential to significantly advance the field of machine learning and enable the development of more advanced and effective models. Its ability to enable continuous learning and adaptability in machine learning models has the potential to revolutionize a wide range of industries and applications.

//Programming Languages of AI

There are many **programming languages** that are commonly used in the field of **artificial intelligence (AI)**.

Some of the most popular programming languages for AI include:

1] Python:
Python is a popular programming language for AI due to its simplicity and ease of use, as well as the large number of libraries and frameworks available for AI tasks such as machine learning and natural language processing.

2] R:
R is a programming language that is commonly used for statistical analysis and data visualization, and is often used in conjunction with machine learning algorithms for AI tasks.

3] Java:
Java is a popular programming language that is widely used in AI due to its portability and the large number of libraries and frameworks available for AI tasks.

4] C++:
C++ is a high-performance programming language that is often used in AI due to its ability to handle large datasets and perform complex calculations efficiently.

5] Lisp:

Lisp is a programming language that was originally developed for AI research and has a long history in the field. It is known for its powerful symbolic processing capabilities and is still used in some areas of AI research.

6] Prolog:

Prolog is a programming language that is well-suited for AI tasks such as natural language processing and symbolic reasoning. It is commonly used in AI research and development.

Overall, there are many programming languages that are commonly used in AI, and the choice of language often depends on the specific AI tasks and the requirements of the project. Some languages may be better suited for certain tasks or environments, while others may offer a more general-purpose approach to AI development.

{ Python

Python is a popular, high-level programming language known for its simplicity, readability, and flexibility. It is an interpreted language, meaning that it is executed at runtime rather than being compiled into machine code. This makes it easier to develop and debug Python programs, as well as making it easier to learn and use for beginners.

Python was first released in 1991 and has since become one of the most popular programming languages in the world. It is used in a wide range of applications, including web development, scientific computing, data analysis, and artificial intelligence (AI).

One of the key features of Python is its large standard library, which provides a wide range of built-in functions and modules for tasks such as connecting to web servers, reading and writing files, and working with data. In addition to the standard library, Python has an active ecosystem of third-party libraries and frameworks that make it even more powerful and flexible.

One of the main reasons for Python's popularity is its simplicity and ease of use. Python has a relatively simple and consistent syntax, making it easy to learn and use for beginners. It also has a large and active community of users and developers, who contribute to the development of the language and its ecosystem of libraries and frameworks.

In addition to its simplicity and flexibility, Python is known for its performance. While it is not the fastest programming language, it is still able to perform complex tasks efficiently due to its efficient implementation and the use of optimized libraries and frameworks.

Overall, Python is a popular and powerful programming language that is widely used in a variety of applications due to its simplicity, flexibility, and performance. Its large standard library and active ecosystem of third-party libraries and frameworks make it an ideal choice for tasks such as web development, scientific computing, data analysis, and artificial intelligence.

Here are some example codes in Python:

1] Printing a message to the console:

```python
print("Hello, World!")
```

2] Declaring and initializing a variable:

```python
x = 10
```

3] Performing arithmetic operations:

```python
y = x + 5
z = x * y
```

4] Creating a function:

```python
def greet(name):
    print("Hello, " + name)

greet("John")
```

5] Creating a loop:

```python
for i in range(10):
    print(i)
```

6] Creating a list:

```python
numbers = [1, 2, 3, 4, 5]
```

7] Accessing elements in a list:

```
print(numbers[0])
print(numbers[-1])
```

8] Modifying elements in a list:

```
numbers[0] = 10
numbers[-1] = 100
```

9] Importing a module:

```
import math

x = math.pi
```

10] Handling exceptions:

```
try:
    x = int("foo")
except ValueError:
    print("Invalid input")
```

"These are just a few examples of the types of code that can be written in Python. Python has a wide range of features and capabilities, making it a powerful and versatile programming language."

{ R

R is a programming language and free software environment for statistical computing and graphics. It is widely used by statisticians, data scientists, and researchers for data analysis, machine learning, and data visualization.

R was developed in the 1990s by Ross Ihaka and Robert Gentleman at the University of Auckland in New Zealand. It is an open-source language, meaning that its source code is freely available and can be modified and distributed by anyone.

One of the key features of R is its large collection of built-in functions and libraries for statistical analysis and data visualization. R also has a wide range of third-party libraries and packages that can be used to extend its capabilities.

R is a high-level language, meaning that it is easy to read and write compared to lower-level languages such as C or C++. It has a syntax that is similar to the S programming language, which was developed at Bell Labs in the 1970s.

R is often used for tasks such as data manipulation, statistical modeling, data visualization, and machine learning. It is particularly well-suited for working with large datasets and performing complex statistical analyses.

R has a large and active community of users and developers, who contribute to the development of the language and its ecosystem of libraries and packages. There are also many resources available for learning R, including online tutorials, textbooks, and courses.

Overall, R is a powerful and widely used programming language and environment for statistical computing and data analysis. Its large collection of built-in functions and libraries, as well as its active community of users and developers, make it an ideal choice for tasks such as data manipulation, statistical modeling, data visualization, and machine learning.

Here are some example codes in R:

1] Printing a message to the console:

```r
print("Hello, World!")
```

2] Declaring and initializing a variable:

```r
x <- 10
```

3] Performing arithmetic operations:

```r
y <- x + 5
z <- x * y
```

4] Creating a function:

```r
greet <- function(name) {
  print(paste("Hello, ", name))
}
greet("John")
```

5] Creating a loop:

```r
for (i in 1:10) {
  print(i)
}
```

6] Creating a vector:

```r
numbers <- c(1, 2, 3, 4, 5)
```

7] Accessing elements in a vector:

```r
print(numbers[1])
print(numbers[length(numbers)])
```

8] Modifying elements in a vector:

```r
numbers[1] <- 10
numbers[length(numbers)] <- 100
```

9] Importing a library:

```r
library(dplyr)

df <- tibble(x = c(1, 2, 3), y = c(4, 5, 6))
df <- df %>% filter(x > 1)
```

10] Handling exceptions:

```
tryCatch({
  x <- as.integer("foo")
}, error = function(e) {
  print("Invalid input")
})
```

"These are just a few examples of the types of code that can be written in R. R has a wide range of features and capabilities, making it a powerful and versatile programming language for statistical computing and data analysis."

{ Java

Java is a popular programming language and computing platform that was first released by Sun Microsystems in 1995. It is an object-oriented language, which means that it is based on the concept of "objects", which can represent real-world entities and their properties and behaviors.

Java is widely used for building applications for the web, mobile devices, and desktop computers. It is also a popular choice for developing Android mobile applications.

One of the key features of Java is its "write once, run anywhere" (WORA) capability, which means that Java code can be run on any device that has a Java Virtual Machine (JVM) installed. This makes Java a portable and platform-independent language.

Java is a statically-typed language, which means that variables have to be declared with a specific type before they can be used. It also has a strong type system, which helps to prevent common programming errors and makes the code more reliable.

Java has a large and active community of users and developers, who contribute to the development of the language and its ecosystem of libraries and frameworks. There are also many resources available for learning Java, including online tutorials, textbooks, and courses.

Overall, Java is a powerful and widely used programming language and computing platform. Its "write once, run anywhere" capability, strong type system, and large community of users and developers make it an ideal choice for building applications for the web, mobile devices, and desktop computers.

Here are some example codes in Java:

1] Printing a message to the console:

```java
System.out.println("Hello, World!");
```

2] Declaring and initializing a variable:

```java
int x = 10;
```

3] Performing arithmetic operations:

```java
int y = x + 5;
int z = x * y;
```

4] Creating a class:

```java
public class Main {
  public static void main(String[] args)
  {
    System.out.println("Hello, World!");
  }
}
```

5] Creating a method:

```java
public class Main {
  public static void greet(String name)
  {
        System.out.println("Hello, " + name);
  }

  public static void main(String[] args)
  {
    greet("John");
  }
}
```

6] Creating a loop:

```java
for (int i = 0; i < 10; i++) {
  System.out.println(i);
}
```

7] Creating an array:

```java
int[] numbers = {1, 2, 3, 4, 5};
```

8] Accessing elements in an array:

```java
System.out.println(numbers[0]);
System.out.println(numbers[numbers.length - 1]);
```

9] Modifying elements in an array:

```java
numbers[0] = 10;
numbers[numbers.length - 1] = 100;
```

10] Handling exceptions:

```java
try {
  int x = Integer.parseInt("foo");
} catch (NumberFormatException e) {
  System.out.println("Invalid input");
}
```

"These are just a few examples of the types of code that can be written in Java. Java has a wide range of features and capabilities, making it a powerful and versatile programming language for building applications for the web, mobile devices, and desktop computers."

{ C++

C++ is a high-performance programming language that was developed by Bjarne Stroustrup in 1979 as an extension of the C programming language. It is a statically-typed, compiled language that is widely used for building large-scale applications, such as operating systems, web browsers, and games.

One of the key features of C++ is its support for object-oriented programming (OOP). This means that C++ allows developers to create "objects" that represent real-world entities and their properties and behaviors. Objects can be created from "classes", which define the characteristics and behaviors of the objects. C++ also supports other programming paradigms, such as procedural programming and generic programming.

C++ is a compiled language, which means that the source code is transformed into machine code that can be executed directly by the computer's processor. This makes C++ programs faster and more efficient than interpreted languages, which have to be translated into machine code at runtime. However, it also means that C++ programs have to be compiled specifically for the target platform, which can be time-consuming.

C++ has a large and active community of users and developers, who contribute to the development of the language and its ecosystem of libraries and frameworks. There are also many resources available for learning C++, including online tutorials, textbooks, and courses.

Overall, C++ is a powerful and widely used programming language that is well-suited for building large-scale applications that require high performance and efficiency. Its support for object-oriented programming and other programming paradigms, as well as its large community of users and developers, make it a popular choice for building applications for a variety of platforms.

Here are some example codes in C++:

1] Printing a message to the console:

```cpp
#include <iostream>

int main() {
    std::cout << "Hello, World!" << std::endl;
    return 0;
}
```

2] Declaring and initializing a variable:

```cpp
int x = 10;
```

3] Performing arithmetic operations:

```
int y = x + 5;
int z = x * y;
```

4] Creating a class:

```
class Point {
 public:
   int x;
   int y;

   Point(int x, int y) {
     this->x = x;
     this->y = y;
   }
};
```

5] Creating a method:

```
class Point {
 public:
   int x;
   int y;

   Point(int x, int y) {
     this->x = x;
     this->y = y;
```

```cpp
    }

    int distance() {
      return sqrt(x * x + y * y);
    }
};
```

6] Creating a loop:

```cpp
for (int i = 0; i < 10; i++) {
  std::cout << i << std::endl;
}
```

7] Creating an array:

```cpp
int[] numbers = {1, 2, 3, 4, 5};
```

8] Accessing elements in an array:

```cpp
std::cout << numbers[0] << std::endl;
std::cout << numbers[sizeof(numbers) / sizeof(int) - 1] << std::endl;
```

9] Modifying elements in an array:

```cpp
numbers[0] = 10;
numbers[sizeof(numbers) / sizeof(int) - 1] = 100;
```

10] Handling exceptions:

```
try {
  int x = std::stoi("foo");
} catch (std::invalid_argument e) {
    std::cout << "Invalid input" << std::endl;
}
```

```
"These are just a few examples of the types
of code that can be written in C++. C++ has
a wide range of features and capabilities,
making it a powerful and versatile
programming language for building
applications for a variety of platforms."
```

{ Lisp

Lisp (short for **"List Processing"**) is a family of programming languages that are characterized by their use of lists as a fundamental data type and their support for symbolic programming. Lisp was first developed in the late 1950s by John McCarthy, and it has since evolved into a number of dialects, including Common Lisp and Scheme.

One of the key features of Lisp is its use of lists as a fundamental data type. Lists are sequences of values that are enclosed in parentheses and separated by spaces. Lists can contain any type of value, including numbers, strings, symbols, and other lists. Lisp also has a rich set of functions and operators for manipulating lists, such as map, filter, and reduce.

Lisp is also known for its support for symbolic programming, which is a style of programming that uses symbols and expressions to represent data and operations. Symbolic programming allows developers to write code that is more flexible and expressive, as it can be easily modified and extended using symbolic expressions. Lisp also has a macro system that allows developers to define their own syntax and control structures, making it a powerful language for building domain-specific languages and customizing the language to fit the needs of a particular problem domain.

Lisp is a dynamically-typed language, which means that the type of a value is determined at runtime rather than at compile time. This makes it easier for developers to write code that is flexible and adaptable, as it can handle a wide range of data types and values. Lisp also has a garbage collector, which automatically manages the memory used by the program and frees up resources that are no longer needed.

Lisp has a long history and a dedicated community of users and developers. It has been used in a wide range of applications, including artificial intelligence, natural language processing, and computer-aided design. Despite its age, Lisp remains a popular and influential programming language that is used in many areas of computer science and industry.

Here are some examples of Lisp code:

1] Printing a message to the console:

```
(print "Hello, World!")
```

2] Declaring and initializing a variable:

```
(setf x 10)
```

3] Performing arithmetic operations:

```
(setf y (+ x 5))
(setf z (* x y))
```

4] Creating a function:

```
(defun add (x y)
   (+ x y))
```

5] Calling a function:

```
(add 3 4)
```

6] Creating a loop:

```
(dotimes (i 10)
   (print i))
```

7] Creating a list:

```
'(1 2 3 4 5)
```

8] Accessing elements in a list:

```
(first '(1 2 3 4 5))
(last '(1 2 3 4 5))
```

9] Modifying elements in a list:

```
(setf (first '(1 2 3 4 5)) 10)
(setf (last '(1 2 3 4 5)) 100)
```

10] Handling exceptions:

```
(handler-case
    (divide 1 0)
  (division-by-zero ()
    (print "Division by zero")))
```

"These are just a few examples of the types of code that can be written in Lisp. Lisp is a flexible and expressive language that is well-suited for a wide range of applications, including artificial intelligence, data processing, and symbolic programming."

{ Prolog

Prolog (short for **"Programming in Logic"**) is a programming language that is based on the principles of logic and symbolic reasoning. Prolog was developed in the 1970s by a group of researchers at the University of Marseille, and it has since become a popular language for artificial intelligence, natural language processing, and symbolic computing.

One of the key features of Prolog is its use of logical statements and rules to represent knowledge and solve problems. In Prolog, a program is a collection of facts and rules that describe the relationships between different entities and concepts. Prolog uses a system of resolution to search for solutions to problems by applying these rules and facts to a given set of data.

Prolog is a declarative language, which means that developers specify what they want the program to do rather than describing how to do it. This allows Prolog to automatically generate efficient search algorithms and solve problems in a more flexible and intuitive way. Prolog also has a built-in unification mechanism that allows it to automatically match and combine data and rules, making it well-suited for tasks that involve pattern matching and data manipulation.

Prolog is often used for artificial intelligence applications, such as natural language processing, expert systems, and planning. It is also used in other areas of computer science, such as database management, automated theorem proving, and symbolic computing. Prolog has a large and active community of users and developers, and it is supported by a wide range of tools and libraries.

Here are some examples of Prolog code:

1] Declaring a fact:

```
likes(alice, apples).
```

2] Declaring a rule:

```
eats(X, Y) :- likes(X, Y).
```

3] Querying the program:

```
?- eats(alice, apples).
```

4] Declaring a list:

```
fruits([apples, bananas, oranges]).
```

5] Accessing elements in a list:

```
head([H|_], H).
tail([_|T], T).
```

6] Recursive function:

```prolog
sum_list([], 0).
sum_list([H|T], N) :- sum_list(T, N1), N is N1 + H.
```

"These are just a few examples of the types of code that can be written in Prolog. Prolog is a powerful and expressive language that is well-suited for a wide range of applications, including artificial intelligence, natural language processing, and symbolic computing."

//Applications of AI

Artificial intelligence (AI) has a wide range of potential applications and is being used in a variety of fields, including:

Healthcare:

AI is being used in healthcare to improve patient care and to help with tasks such as diagnosis, treatment planning, and medication management.

Finance:

AI is being used in finance to help with tasks such as fraud detection, credit risk analysis, and portfolio management.

Retail:

AI is being used in retail to help with tasks such as demand forecasting, personalization, and inventory management.

Manufacturing:

AI is being used in manufacturing to improve efficiency, quality, and safety, and to help with tasks such as inspection, maintenance, and production planning.

Transportation:

AI is being used in transportation to help with tasks such as autonomous driving, traffic management, and logistics.

Education:

AI is being used in education to help with tasks such as personalized learning, assessment, and tutoring.

Agriculture:

AI is being used in agriculture to help with tasks such as precision farming, crop monitoring, and irrigation management.

Robotics:

AI is being used in robotics to help with tasks such as perception, navigation, and manipulation.

AI has a wide range of potential applications and is being used in a variety of fields to improve efficiency, productivity, and quality. It is a rapidly growing field with the potential to transform the way humans and computers interact with the world around them. Let's take a deep look few among those applications in details.

{ Healthcare

Artificial intelligence (AI) has the potential to revolutionize the healthcare industry by enhancing patient care and increasing efficiency. Here are a few examples of how AI is being used in healthcare:

1] Diagnosis:

AI algorithms can analyze medical images, such as x-rays and CT scans, and process vast amounts of patient data to identify patterns and trends that may be difficult for humans to discern. This can help with early detection and diagnosis of diseases.

2] Treatment planning:

AI can analyze patient data and medical knowledge to suggest the most effective treatment options based on a patient's specific needs and circumstances.

3] Medication management:

AI can help identify potential drug interactions and suggest appropriate dosages by analyzing patient data and medical knowledge.

4] Clinical decision support:

AI can provide real-time guidance and recommendations to doctors based on a patient's medical history and current condition.

5] Population health management:

By analyzing data from large numbers of patients, AI can identify trends and patterns that may help to improve the overall health of a population.

6] Clinical trial matching:

AI can help match patients to clinical trials by analyzing patient data and medical knowledge to identify suitable candidates.

AI has the potential to transform healthcare by improving patient care, increasing efficiency, and reducing costs. However, it is important to also consider the challenges and concerns associated with the use of AI in healthcare, such as issues related to privacy, bias, and liability.

Also artificial intelligence (AI) is being used in a number of healthcare applications today, and the number is likely to continue to grow in the coming years as the technology continues to advance. Here are a few examples of actual healthcare applications that are using AI:

Enlitic: Enlitic is a company that uses deep learning algorithms to analyze medical images, such as x-rays and CT scans, to help with diagnosis and treatment planning.

MedyMatch: MedyMatch is a company that uses AI to analyze patient data and medical knowledge to help with diagnosis, treatment planning, and medication management.

Sensely: Sensely is a company that uses AI to provide virtual nursing assistants to help with patient triage, symptom checking, and health tracking.

Atomwise: Atomwise is a company that uses AI to help with drug discovery and development by analyzing large amounts of data to identify potential new drug candidates.

CloudMedx: CloudMedx is a company that uses AI to provide real-time clinical decision support to doctors by analyzing patient data and medical knowledge.

There are already a number of healthcare applications that are using AI, and the number is likely to continue to grow in the coming years as the technology continues to advance.

{ Finance

Finance is a broad term that refers to the management of money and assets. It includes the creation, distribution, and management of financial products and services, such as loans, investments, and insurance. Finance plays a crucial role in the economy by facilitating the exchange of money and helping to allocate resources effectively.

There are several subfields within finance, including personal finance, corporate finance, and public finance. Personal finance involves managing an individual's financial resources, such as their income, expenses, and investments. Corporate finance involves managing the financial resources of a business, including the raising of capital and the allocation of financial resources. Public finance involves managing the financial resources of a government, including the collection of taxes and the allocation of public funds.

Overall, finance plays a crucial role in the economy by facilitating the exchange of money and helping to allocate resources effectively. It is a complex and dynamic field that involves the management of financial resources at all levels, from individuals to businesses to governments.

Artificial intelligence (AI) has the potential to revolutionize the finance industry by improving efficiency, reducing costs, and enabling more effective decision making. Here are a few examples of how AI is contributing to the finance industry:

1] Fraud detection:

AI algorithms can analyze large amounts of data to identify patterns and anomalies that may indicate fraudulent activity, helping to reduce the risk of financial losses due to fraud.

2] Credit risk analysis:

AI can analyze a borrower's financial data and predict the likelihood of default, helping financial institutions to make more informed lending decisions and to manage risk.

3] Portfolio management:

AI can analyze market data and make investment decisions based on that analysis, helping to optimize portfolio performance.

4] Trading:

AI can analyze market data and make trades based on that analysis, helping to improve the speed and efficiency of trading operations.

5] Customer service:

AI-powered chatbots and other automated systems can provide real-time assistance to customers, helping to improve the customer experience and reduce the burden on human customer service staff.

6] Risk management:

AI can analyze data and make predictions about potential risks that a financial institution may face, helping to identify and mitigate potential risks.

AI has the potential to transform the finance industry by improving efficiency, reducing costs, and enabling more effective decision making. However, there are also challenges and concerns associated with the use of AI in finance, including issues related to bias and accountability, that will need to be addressed.

Here are a few examples of how artificial intelligence (AI) is being used in the finance industry:

Capital One: Capital One is a financial services company that is using AI to help with fraud detection by analyzing large amounts of data to identify patterns and anomalies that may indicate fraudulent activity.

FICO: FICO is a credit scoring company that is using AI to help with credit risk analysis by analyzing a borrower's financial data and predicting the likelihood of default.

BlackRock: BlackRock is a financial services company that is using AI to help with portfolio management by analyzing market data and making investment decisions based on that analysis.

JP Morgan: JP Morgan is a financial services company that is using AI to help with trading by analyzing market data and making trades based on that analysis.

American Express: American Express is a financial services company that is using AI-powered chatbots to provide real-time assistance to customers through its website and mobile app.

AIG: AIG is an insurance company that is using AI to help with risk management by analyzing data and making predictions about potential risks that the company may face.

Overall, these are just a few examples of how AI is being used in the finance industry to improve efficiency, reduce costs, and enable more effective decision making. There are many other examples of AI being used in finance, and the number is likely to continue to grow in the coming years as the technology continues to advance.

{ Transportation

Transportation refers to the movement of people, goods, and other materials from one place to another. It plays a crucial role in the economy by enabling the exchange of goods and services and by facilitating the movement of people. There are many different modes of transportation, including air, land, and water, and each mode has its own set of advantages and disadvantages.

In the land transportation category, there are several sub-modes including cars, buses, trains, and bicycles. Air transportation includes planes and helicopters, while water transportation includes ships, boats, and ferries. Public transportation, such as buses and trains, is a common way for people to get around in many cities, while personal transportation, such as cars and bikes, is more common in suburban and rural areas.

Overall, transportation plays a vital role in the economy by enabling the exchange of goods and services and facilitating the movement of people. It is a complex and dynamic field that involves the movement of people, goods, and other materials from one place to another using various modes of transportation.

Artificial intelligence (AI) has the potential to transform the transportation industry by improving safety, efficiency, and convenience. Here are a few examples of how AI is contributing to the transportation industry:

1] Autonomous vehicles:

AI is being used to develop autonomous vehicles, such as self-driving cars and trucks, which are capable of navigating roads and highways without the need for human intervention.

2] Traffic management:

AI can be used to analyze traffic data and make predictions about traffic flow, helping to optimize routes and reduce congestion.

3] Public transportation:

AI can be used to optimize schedules and routes for public transportation, such as buses and trains, helping to improve efficiency and reduce costs.

4] Logistics:

AI can be used to optimize logistics operations, such as the delivery of goods, by analyzing data and making predictions about demand and supply.

5] Personalized travel recommendations:

AI can be used to provide personalized travel recommendations to users based on their past travel history and preferences.

6] Predictive maintenance:

AI can be used to predict when equipment, such as planes and trains, is likely to fail, helping to reduce downtime and improve safety.

Overall, AI has the potential to transform the transportation industry by improving safety, efficiency, and convenience. However, there are also challenges and concerns associated with the use of AI in transportation, including issues related to liability and regulation, that will need to be addressed.

Here are a few examples of how artificial intelligence (AI) is being used in the transportation industry:

Waymo: Waymo is a company that is using AI to develop autonomous vehicles, such as self-driving cars and trucks.

Waze: Waze is a navigation app that uses AI to analyze traffic data and make predictions about traffic flow, helping to optimize routes and reduce congestion.

Citymapper: Citymapper is a public transportation app that uses AI to optimize schedules and routes for buses and trains, helping to improve efficiency and reduce costs.

Convoy: Convoy is a logistics company that uses AI to optimize the delivery of goods by analyzing data and making predictions about demand and supply.

Hopper: Hopper is a travel app that uses AI to provide personalized travel recommendations to users based on their past travel history and preferences.

GE Aviation: GE Aviation is a company that uses AI to predict when equipment, such as planes and trains, is likely to fail, helping to reduce downtime and improve safety.

Overall, these are just a few examples of how AI is being used in the transportation industry to improve safety, efficiency, and convenience. There are many other examples of AI being used in transportation, and the number is likely to continue to grow in the coming years as the technology continues to advance.

{ Customer Service

Customer service refers to the support and assistance provided to customers before, during, and after a purchase. It plays a crucial role in the customer experience, as it can help to build trust, loyalty, and satisfaction. There are many different channels through which customer service can be provided, including phone, email, chat, and social media.

Good customer service should be timely, efficient, and helpful, and it should aim to resolve any issues or concerns that a customer may have. It can involve answering questions, providing information, processing orders and returns, and handling complaints and feedback.

In the digital age, many companies are using artificial intelligence (AI) to help with customer service, such as chatbots and virtual assistants. These AI-powered tools can provide real-time assistance to customers, helping to improve the customer experience and reduce the burden on human customer service staff.

Customer service plays a crucial role in the customer experience, as it can help to build trust, loyalty, and satisfaction. It is an important part of any business, and it involves providing support and assistance to customers before, during, and after a purchase.

Artificial intelligence (AI) has the potential to transform the customer service industry by improving efficiency, accuracy, and personalization. Here are a few examples of how AI is contributing to the customer service industry:

1] Chatbots:
AI-powered chatbots can provide real-time assistance to customers through websites and mobile apps, answering questions, providing information, and helping to resolve issues.

2] Virtual assistants:
AI-powered virtual assistants, such as Apple's Siri and Amazon's Alexa, can provide personalized assistance to users, including

answering questions, providing information, and helping to complete tasks.

3] Customer service automation:

AI can be used to automate certain aspects of customer service, such as routing calls and processing orders and returns, helping to improve efficiency and reduce the burden on human customer service staff.

4] Sentiment analysis:

AI can be used to analyze customer feedback and identify patterns and trends, helping to identify areas for improvement and to tailor the customer experience to individual needs.

5] Personalization:

AI can be used to provide personalized recommendations and experiences to customers based on their past interactions and preferences.

AI has the potential to transform the customer service industry by improving efficiency, accuracy, and personalization. However, there are also challenges and concerns associated with the use of AI in customer service, including issues related to bias and accountability, that will need to be addressed.

Here are a few examples of how artificial intelligence (AI) is being used in the customer service industry:

Uber: Uber uses AI-powered chatbots to provide real-time assistance to customers through its mobile app, answering questions and helping to resolve issues.

Apple: Apple's virtual assistant, Siri, uses AI to provide personalized assistance to users, including answering questions, providing information, and helping to complete tasks.

Amazon: Amazon's virtual assistant, Alexa, uses AI to provide personalized assistance to users, including answering questions, providing information, and helping to complete tasks.

Zendesk: Zendesk is a customer service software company that uses AI to automate certain aspects of customer service, such as routing calls and processing orders and returns, helping to improve efficiency and reduce the burden on human customer service staff.

Salesforce: Salesforce is a customer relationship management company that uses AI to analyze customer feedback and identify patterns and trends, helping to identify areas for improvement and to tailor the customer experience to individual needs.

Netflix: Netflix uses AI to provide personalized recommendations to users based on their past viewing history and preferences.

Overall, these are just a few examples of how AI is being used in the customer service industry to improve efficiency, accuracy, and personalization. There are many other examples of AI being used in customer service, and the number is likely to continue to grow in the coming years as the technology continues to advance.

//Ethics and Society

Ethics and society in artificial intelligence (AI) refer to the moral and social implications of the development and use of AI. As AI becomes increasingly sophisticated and integrated into various aspects of society, it is important to consider the ethical and social implications of the technology.

Some of the key ethical and social issues related to AI include:

1] Bias:
AI systems can be biased if they are trained on biased data, leading to unfair outcomes. It is important to ensure that AI systems are trained on diverse and representative data to minimize bias.

2] Privacy:
AI systems often require the collection and processing of personal data, raising concerns about privacy. It is important to ensure that AI systems are designed and used in a way that respects individuals' privacy rights.

3] Transparency:

AI systems can be complex and difficult to understand, making it difficult to hold them accountable for their decisions. It is important to ensure that AI systems are transparent and explainable, so that their decisions can be understood and scrutinized.

4] Employment:

AI has the potential to automate many tasks, which could lead to job loss and displacement. It is important to consider the social impacts of AI on employment and to ensure that the benefits of AI are shared equitably.

It is important to consider the ethical and social implications of AI as the technology continues to advance and become more integrated into society. It is crucial to ensure that AI is developed and used in a way that is ethical, responsible, and beneficial to society.

{ Bias in AI

Bias in artificial intelligence (AI) refers to the systematic and unintentional discrimination of certain groups or individuals by AI systems. Bias can occur when AI systems are trained on biased data, which can lead to unfair and unequal outcomes.

There are several sources of bias in AI, including:

1] Training data:

AI systems are trained on large datasets, and if the data is biased, the resulting AI system will also be biased. For example, if a facial recognition system is trained on a dataset that is predominantly white faces, it may be less accurate at recognizing faces from other racial groups.

2] Algorithms:

The algorithms that are used to develop AI systems can also be biased. For example, if an algorithm is designed to optimize for a certain outcome, it may perpetuate biases that are present in the data.

3] Human bias:

Human bias can also be introduced into AI systems through the design and development process. For example, if the designers of an AI system have their own biases, they may inadvertently incorporate those biases into the system.

Bias in AI can have serious consequences, as it can lead to unfair and unequal treatment of certain groups or individuals. It is important to recognize and address bias in AI to ensure that the technology is developed and used in an ethical and responsible manner. This can involve a variety of approaches, such as developing AI systems that are trained on diverse and representative data and designing algorithms that are transparent and explainable.

Bias in artificial intelligence (AI) can have both advantages and disadvantages, depending on the context and the specific consequences of the bias. Here are a few potential advantages and disadvantages of bias in AI:

Advantages:

Efficiency: Bias in AI can sometimes lead to increased efficiency, as the AI system may focus on a specific group or task and be able to optimize for that particular objective.

Profit: Bias in AI can sometimes lead to increased profits, as the AI system may be able to optimize for a specific outcome that is beneficial to the organization.

Disadvantages:

Inequality: Bias in AI can lead to unequal and unfair treatment of certain groups or individuals, which can have serious social and economic consequences.

Legal consequences: Bias in AI can lead to legal consequences, as it can violate laws and regulations that prohibit discrimination.

Loss of trust: Bias in AI can lead to a loss of trust in the technology, as it can be perceived as unfair or unethical.

Overall, bias in AI can have both advantages and disadvantages, depending on the context and the specific consequences of the bias. It is important to recognize and address bias in AI to ensure that the technology is developed and used in an ethical and responsible manner.

{ Job Displacement

Job displacement refers to the loss of jobs due to technological change, such as the adoption of automation or artificial intelligence (AI). As AI becomes increasingly sophisticated and integrated into various industries, it has the potential to automate many tasks, which could lead to job loss and displacement.

There are several factors that can contribute to job displacement due to AI:

Automation: AI has the ability to automate many tasks, which can lead to job loss and displacement. For example, AI-powered robots can perform tasks such as assembly line work and warehousing, which could lead to job loss for human workers.

Productivity: AI can increase productivity and efficiency, which can lead to a reduction in the number of jobs needed to produce a given output. For example, if a company introduces an AI system that can perform a task faster and more accurately than a human worker, the company may decide to reduce the number of human workers.

Changes in the nature of work: AI can also change the nature of work, which can lead to job loss and displacement. For example, if a company introduces an AI system that can perform a task more accurately and consistently than a human worker, the company may decide to reassign the human worker to a different task.

Job displacement due to AI is a complex and multifaceted issue, and it will likely have different impacts on different industries and workers. It is important to consider the potential impacts of AI on employment and to ensure that the benefits of AI are shared equitably.

Job displacement due to artificial intelligence (AI) can have major effects on human workers, including:

1] Unemployment:

Job displacement due to AI can lead to unemployment, as workers may lose their jobs as a result of automation or other technological changes.

2] Wages:

Job displacement due to AI can also affect wages, as workers who are able to find new employment may have to accept lower wages or less favorable working conditions.

3] Inequality:

Job displacement due to AI can contribute to inequality, as certain groups of workers may be more vulnerable to job loss and may have fewer options for finding new employment.

4] Psychological impacts:

Job displacement due to AI can have psychological impacts on workers, such as stress, anxiety, and a sense of loss or uncertainty about the future.

5] Social impacts:

Job displacement due to AI can have social impacts, such as a decline in the number of middle-class jobs and a decline in the purchasing power of workers.

So, job displacement due to AI can have major effects on human workers, including unemployment, wage declines, inequality, and psychological and social impacts. It is important to consider the potential impacts of AI on employment and to ensure that the benefits of AI are shared equitably.

There are several ways that humans can potentially overcome the challenges posed by job displacement due to artificial intelligence (AI). Here are a few potential strategies:

Education and training: One way to overcome job displacement due to AI is to invest in education and training that helps workers acquire new skills and knowledge. This can help workers to adapt to new technologies and find employment in new industries or occupations.

Labor market policies: Governments can also implement labor market policies that help workers to adapt to technological change, such as unemployment insurance, retraining programs, and income support.

Social safety nets: Governments can also put in place social safety nets, such as healthcare, housing, and social security, which can help to cushion the impact of job loss on workers and their families.

Investment in R&D: Governments and businesses can invest in research and development (R&D) to create new technologies and industries that can provide employment opportunities for workers.

Collaboration between humans and AI: Another approach is to focus on ways to collaborate between humans and AI, rather than replacing human workers with AI. This could involve tasks that are complementary to AI, rather than substituting for them.

Overall, there are several strategies that can help humans to overcome the challenges posed by job displacement due to AI. It is important to consider a combination of approaches, as different strategies may be more effective in different contexts.

{ Potential for Misuse

The **potential for misuse** refers to the risk that artificial intelligence (AI) systems may be used in ways that are harmful, unethical, or against the interests of society. There are several potential sources of misuse of AI:

1] Malicious actors:
AI systems can be vulnerable to hacking and other forms of cyberattacks, which could allow malicious actors to access and

misuse sensitive data or manipulate the system for their own benefit.

2] Unethical use:

AI systems can also be used in ways that are unethical or that violate the rights of individuals. For example, an AI system could be used to engage in discrimination or surveillance, or to manipulate public opinion or elections.

3] Unintended consequences:

The use of AI can also have unintended consequences, such as unintended biases or errors that can have negative impacts on individuals or society.

4] Misuse by governments:

Governments can also misuse AI to further their own agendas or to suppress dissent or minority groups.

Overall, the potential for misuse of AI is a serious concern, as it can have serious consequences for individuals and society. It is important to consider the potential risks of AI and to develop strategies to minimize the potential for misuse of the technology. This can involve a variety of approaches, such as ethical guidelines, regulations, and accountability mechanisms.

There are several strategies that can be used to prevent the potential for misuse of artificial intelligence (AI). Here are a few potential approaches:

1] Ethical guidelines:

Developing ethical guidelines for the development and use of AI can help to ensure that the technology is used in a responsible and ethical manner. These guidelines can address issues such as fairness, accountability, transparency, and human rights.

2] Regulations:

Governments can also regulate the development and use of AI, setting standards and rules that must be followed by organizations and individuals. Regulations can address issues such as data privacy, cybersecurity, and the use of AI in sensitive areas such as healthcare and criminal justice.

3] Accountability mechanisms:

It is also important to establish accountability mechanisms that can hold organizations and individuals responsible for the misuse of AI. This can involve approaches such as legal liabilities, oversight committees, and independent audits.

4] Public education and engagement:

Increasing public awareness and understanding of AI can also help to prevent misuse of the technology. This can involve educating the public about the capabilities and limitations of AI, as well as engaging with stakeholders to ensure that their concerns and needs are taken into account.

Overall, there are several strategies that can be used to prevent the potential for misuse of AI. It is important to consider a combination of approaches, as different strategies may be more effective in different contexts.

{ Regulation and Governance

Regulation and governance of artificial intelligence (AI) refers to the policies, laws, and institutions that are used to oversee the development and use of AI. The goal of regulation and governance of AI is to ensure that the technology is developed and used in a responsible and ethical manner, while also fostering innovation and economic growth.

There are several challenges to regulating and governing AI, including:

1] Complexity:
AI is a complex and rapidly evolving technology, which makes it difficult to regulate and govern effectively.

2] Lack of consensus:
There is often a lack of consensus on how to regulate and govern AI, as different stakeholders may have different priorities and perspectives.

3] Jurisdictional issues:
AI can cross national borders, which can create jurisdictional issues and make it difficult to establish consistent regulations and governance frameworks.

4] Difficulties in predicting the future:

It is also difficult to predict the future impacts and implications of AI, which can make it challenging to regulate and govern the technology in a proactive and effective manner.

Overall, regulation and governance of AI is a complex and multifaceted issue, and it will likely involve a combination of approaches, such as ethical guidelines, regulations, accountability mechanisms, and public education and engagement. It is important to ensure that the benefits of AI are shared equitably and that the technology is developed and used in a responsible and ethical manner.

Regulation and governance of artificial intelligence (AI) is important for humans because AI has the potential to impact many aspects of society and to shape the future in significant ways. Here are a few reasons why regulation and governance of AI is important for humans:

Ethical considerations: AI has the potential to raise ethical and moral concerns, such as fairness, accountability, and transparency. Regulation and governance can help to ensure that AI is developed and used in a responsible and ethical manner.

Economic impacts: AI has the potential to disrupt labor markets and to affect the distribution of wealth and income. Regulation and governance can help to ensure that the economic impacts of AI are managed in a way that is fair and equitable.

Social impacts: AI can also have social impacts, such as changing the nature of work and the role of humans in society. Regulation and governance can help to ensure that the social impacts of AI are managed in a way that is beneficial to society.

Privacy and security: AI can also raise privacy and security concerns, as it often involves the processing of large amounts of sensitive data. Regulation and governance can help to ensure that the privacy and security of individuals is protected.

Overall, regulation and governance of AI is important for humans because it can help to ensure that the technology is developed and used in a responsible and ethical manner, and that the benefits and impacts of AI are shared equitably.

//Future of AI

The **future of artificial intelligence (AI)** is difficult to predict with certainty, as it will depend on many factors, including technological advances, economic conditions, and social and political factors. However, it is likely that AI will continue to play an increasingly important role in many aspects of society, including:

1] Automation:
AI has the potential to automate many tasks, which could lead to increased productivity and efficiency. However, it could also lead to job displacement, particularly in industries that are more susceptible to automation.

2] Employment:

The future of employment may be influenced by AI, as the technology has the potential to create new jobs as well as displace existing ones. It is important to consider the potential impacts of AI on employment and to ensure that the benefits of AI are shared equitably.

3] Economic growth:

AI has the potential to drive economic growth through innovation and productivity gains. However, it is important to consider the distribution of the benefits of AI and to ensure that they are shared equitably.

4] Healthcare:

AI has the potential to transform healthcare, through applications such as diagnosis, treatment planning, and drug discovery. However, it is important to consider the ethical and privacy implications of using AI in healthcare.

5] Transportation:

AI has the potential to transform transportation, through the development of autonomous vehicles and other technologies. However, it is important to consider the safety and regulatory implications of using AI in transportation.

Overall, the future of AI is uncertain, but it is likely to play an increasingly important role in many aspects of society. It is important to consider the potential impacts and implications of AI and to ensure that the technology is developed and used in a responsible and ethical manner.

The relationship between artificial intelligence (AI) and humans in the future will depend on many factors, including technological advances, economic conditions, and social and political factors. Here are a few potential scenarios for the relationship between AI and humans in the future:

1] Collaboration:

AI and humans could collaborate in the future, with AI augmenting human capabilities and working alongside humans to achieve common goals. This could involve tasks that are complementary to AI, rather than substituting for humans.

2] Competition:

AI and humans could also compete in the future, as AI may be able to automate certain tasks more effectively or efficiently than humans. This could lead to job displacement and other economic impacts, and it is important to consider the distribution of the benefits of AI and to ensure that they are shared equitably.

3] Coexistence:

AI and humans could also coexist in the future, with AI and humans working in different spheres or sectors. This could involve humans focusing on tasks that are more creative, social, or emotional, while AI handles tasks that are more routine or technical.

Overall, the relationship between AI and humans in the future is uncertain, but it is likely to be complex and multifaceted. It is important to consider the potential impacts and implications of AI and to ensure that the technology is developed and used in a responsible and ethical manner.

{ Potential Advancements

Potential advancements refer to the possible developments or improvements that could be made in a particular field in the future. In the context of artificial intelligence (AI), potential advancements could include:

Enhanced capabilities: AI could continue to become more intelligent and capable over time, with the ability to perform more complex tasks and to learn from new experiences. This could lead to a range of benefits, including increased productivity and efficiency, as well as new applications and innovations.

Human-like intelligence: AI could also potentially achieve human-like intelligence, with the ability to understand and interpret complex situations and to exhibit creativity and independent thought. This could lead to a range of benefits, but it could also raise ethical and social concerns.

Improved decision-making: AI could also potentially improve decision-making, with the ability to analyze large amounts of data and to make more accurate and unbiased decisions. This could lead to a range of benefits, such as better healthcare outcomes, improved transportation safety, and more efficient resource allocation.

Enhanced communication: AI could also potentially improve communication, with the ability to understand and interpret natural language and to interact with humans more effectively. This could lead to a range of benefits, such as improved customer service and more effective collaboration between humans and AI.

Overall, there are many potential advancements that could be made in AI in the future, which could have a wide range of benefits and impacts on society. It is important to consider the potential impacts and implications of AI and to ensure that the technology is developed and used in a responsible and ethical manner.

Artificial intelligence (AI) is a rapidly developing field with the potential to revolutionize the way we live and work. As AI continues to advance, it is likely to bring about a wide range of benefits for humans. Here are a few ways in which AI could potentially improve our lives in the future:

1] Streamlined tasks:

AI has the ability to automate many tasks that are currently performed by humans. This could lead to increased efficiency and productivity, as well as more leisure time for individuals.

2] Improved decision-making:

AI's ability to analyze large amounts of data and make informed decisions could lead to better outcomes in a range of areas, such as healthcare, finance, and transportation.

3] New job opportunities:

As AI continues to advance, it is likely to create new industries and businesses that did not previously exist. This could lead to the creation of new job opportunities for skilled professionals.

4] Enhanced quality of life:

AI has the potential to improve the quality of life for individuals in many ways, such as through personalized healthcare, improved education, and enhanced entertainment experiences.

Overall, AI has the potential to bring about a wide range of benefits for humans. It is important to consider the potential impacts and implications of AI and to ensure that the technology is developed and used in a responsible and ethical manner.

{ Challenges and Limitations

Artificial intelligence (AI) is a rapidly developing field that has the potential to revolutionize many aspects of society. However, there are also many challenges and limitations to the development and use of AI. Here are a few examples:

1] Ethical considerations:

AI has the potential to raise ethical and moral concerns, such as fairness, accountability, and transparency. For example, AI systems that are used to make decisions about hiring, lending, or sentencing may be biased against certain groups, leading to unfair or discriminatory outcomes. Ensuring that AI is developed and used in a responsible and ethical manner is an ongoing challenge.

2] Economic impacts:

AI has the potential to disrupt labor markets and to affect the distribution of wealth and income. As AI automates certain tasks, it may lead to job displacement, which could have negative economic impacts on certain workers. Ensuring that the economic impacts of AI are managed in a way that is fair and equitable is an ongoing challenge.

3] Social impacts:

AI can also have social impacts, such as changing the nature of work and the role of humans in society. For example, AI may lead to changes in the skills that are in demand, or it may alter the balance of power between workers and employers. Ensuring that the social impacts of AI are managed in a way that is beneficial to society is an ongoing challenge.

4] Privacy and security:

AI often involves the processing of large amounts of sensitive data, which raises privacy and security concerns. For example, AI systems that are used to analyze medical records or personal financial data must be designed to protect the privacy of individuals. Protecting the privacy and security of individuals is an ongoing challenge.

5] Bias:

AI systems can be biased, either intentionally or unintentionally, which can lead to unfair or discriminatory outcomes. For example, AI systems may be trained on biased data, which could lead to biased results. Reducing bias in AI systems is an ongoing challenge.

Overall, there are many challenges and limitations to the development and use of AI. It is important to consider these challenges and to work towards addressing them in order to ensure that the benefits of AI are shared equitably and that the technology is developed and used in a responsible and ethical manner.

{ Implications for Society

The implications of artificial intelligence (AI) for society are likely to be wide-ranging and complex. Some of the potential implications of AI include:

1] Changes in the nature of work:

AI has the potential to automate many tasks that are currently performed by humans, which could lead to changes in the nature of work. This could result in job displacement for some workers, but it could also create new job opportunities in fields such as AI development and maintenance.

2] Changes in the distribution of wealth and income:

AI has the potential to affect the distribution of wealth and income, as it may lead to changes in the demand for certain skills and the value of labor. Ensuring that the economic impacts of AI are managed in a way that is fair and equitable is an ongoing challenge.

3] Changes in social relationships and interactions:

AI has the potential to change the way that people interact and relate to one another, as it may alter the balance of power between individuals and groups. It is important to consider the social implications of AI and to ensure that the technology is used in a way that is beneficial to society.

4] Changes in the way that we access and consume information:

AI has the potential to change the way that we access and consume information, as it may enable the development of more personalized and targeted content. This could have implications for the media industry and for the way that we consume news and entertainment.

5] Changes in the way that we live:

AI has the potential to change the way that we live in many other ways as well. For example, it may enable the development of new technologies and products, such as autonomous vehicles, personalized healthcare, and intelligent home appliances.

Overall, the implications of AI for society are likely to be complex and far-reaching. It is important to consider the potential impacts of AI and to ensure that the technology is developed and used in a responsible and ethical manner.

//Conclusion

Artificial intelligence (AI) is a rapidly developing field that has the potential to transform many aspects of society. It has the potential to change the nature of work, the distribution of wealth and income, social relationships, and the way that we access and consume information. In many cases, AI has the potential to improve efficiency, accuracy, and accessibility, making it possible to solve complex problems and to address social and economic challenges in new ways.

However, the development and use of AI also raise many challenges and limitations. One key challenge is the ethical implications of AI. As AI systems become more sophisticated and are used to make decisions that affect people's lives, it is important to ensure that they are developed and used in a responsible and ethical manner. This may require the development of new regulations, standards, and oversight mechanisms to ensure that AI is used in a way that is fair, transparent, and accountable.

Another challenge is the economic impacts of AI. As AI automates certain tasks, it has the potential to disrupt labor markets and to affect the distribution of wealth and income. Ensuring that the economic impacts of AI are managed in a way that is fair and equitable is an ongoing challenge. This may require the development of new policies and programs to support workers who are affected by AI-driven job displacement, as well as efforts to encourage the development of new job opportunities in fields such as AI development and maintenance.

AI also has the potential to have social impacts, such as changing the nature of work and the role of humans in society. Ensuring that the social impacts of AI are managed in a way that is beneficial to society is an ongoing challenge. This may require the development of new policies and programs to support workers who are affected by AI-driven job displacement, as well as efforts to encourage the development of new job opportunities in fields such as AI development and maintenance.

Privacy and security are also key concerns with AI. As AI systems process large amounts of sensitive data, it is important to ensure that the privacy and security of individuals are protected. This may require the development of new technologies and regulations to protect personal data and to ensure that AI systems are secure from cyber attacks and other threats.

Bias is another key challenge with AI. AI systems can be biased, either intentionally or unintentionally, which can lead to unfair or discriminatory outcomes. Reducing bias in AI systems is an ongoing challenge. This may require the development of new methods and tools to identify and mitigate bias in AI systems, as well as efforts to ensure that AI is developed and used in a way that is transparent and accountable.

Overall, AI has the potential to transform many aspects of society, but it also raises many challenges and limitations. It is important to consider these challenges and limitations and to work towards addressing them in order to ensure that the benefits of AI are shared equitably and that the technology is developed and used in a responsible and ethical manner. This may require the development of new policies, regulations, and oversight mechanisms to ensure that AI is used in a way that is fair, transparent, and accountable. It may also require the development of new technologies and approaches to address issues such as bias, privacy and security, and economic impacts. By addressing these challenges and limitations, it is possible to ensure that AI is developed and used in a way that benefits society as a whole.

{ Summary of Key Points

Artificial intelligence (AI) has the potential to significantly impact various aspects of society, from the way we work and interact to the distribution of wealth and the access and consumption of information. While AI offers numerous benefits such as increased efficiency and accuracy, it also poses various challenges that need to be addressed. Ethical considerations, economic impacts, social impacts, privacy and security concerns, and bias are just a few of the issues that need to be considered when developing and implementing AI. To ensure that the benefits of AI are shared fairly and that the technology is used responsibly, it is necessary to establish regulations, standards, and oversight mechanisms. Additionally, it is crucial to address the potential displacement of jobs by supporting affected workers and encouraging the creation of new job opportunities in the AI field. Ensuring that AI is developed and used in a transparent and accountable manner is also important, as is developing methods and tools to identify and mitigate bias in AI systems. By addressing these challenges and limitations, we can work towards a future where AI benefits society as a whole.

Here are some key points that can be drawn from the information provided:

1] Artificial intelligence (AI) is a rapidly developing field with the potential to transform many aspects of society.

2] While AI offers many benefits, it also poses various challenges and limitations such as ethical considerations, economic impacts, social impacts, privacy and security concerns, and bias.

3] It is important to consider these challenges and limitations and to work towards addressing them in order to ensure that the benefits of AI are shared equitably and that the technology is used responsibly.

4] This may require the development of new policies, regulations, and oversight mechanisms, as well as new technologies and approaches to address issues such as bias, privacy and security, and economic impacts.

5] The future of AI has the potential to significantly impact various aspects of society, including work, social relationships, and the way we access and consume information.

6] Ensuring that the benefits of AI are shared fairly and that the technology is used responsibly will require addressing challenges and limitations such as ethical considerations, economic impacts, social impacts, privacy and security concerns, and bias.

{ Future Directions for AI Research and Development

The future direction of AI research and development is likely to focus on a range of areas, including artificial general intelligence, AI safety, explainability, and more specialized areas such as natural language processing, machine learning, computer vision, and robotics.

Artificial general intelligence (AGI) involves the development of AI systems that have the ability to perform a wide range of tasks at the same level of human intelligence. This is a challenging and ambitious goal, but if achieved it could have significant implications for society and the way we live and work. For example, AGI could potentially be used to solve complex problems that currently require human expertise, such as finding a cure for a particular disease or developing a new energy source. It could also be used to automate certain tasks, potentially leading to increased efficiency and productivity. However, the development of AGI also raises ethical and social concerns, such as the potential displacement of jobs and the risk of AI systems acting in ways that are harmful to humans. Ensuring that AGI is developed and used in a responsible and ethical manner will be an important area of focus for AI research and development in the future.

AI safety is another important area of focus for the future of AI research and development. This involves developing ways to ensure that AI systems are safe and reliable, and that they behave in a manner that is beneficial to humans. This includes developing ways to prevent AI systems from causing harm, such as through the use of safety constraints and monitoring systems, as well as methods for controlling and mitigating the risks of AI. Ensuring the safety of AI systems will be particularly important as they become more widespread and are used in critical applications such as healthcare, transportation, and national defense.

Explainability is another key area of focus for the future of AI research and development. This involves developing AI systems that are able to explain their decision-making processes and reasoning to humans. This is important for ensuring that AI systems are transparent and accountable, and for building trust in the technology. For example, if an AI system is used to make a decision that affects a person's life, such as whether or not to grant them a loan or to hire them for a job, it is important that the person understands the reasoning behind the decision. Explainability is also important for helping to identify and mitigate bias in AI systems, as it allows humans to understand how the system arrived at a particular decision and to determine if the decision was based on biased data or algorithms.

Other areas of focus for AI research and development include natural language processing, machine learning, computer vision, and robotics. These areas are likely to see significant advances in the coming years, with the potential to have a wide range of applications in fields such as healthcare, transportation, and customer service. For example, natural language processing could be used to develop virtual assistants that are able to understand and respond to human voice commands, or to develop machine translation systems that are able to translate text or speech from one language to another. Machine learning could be used to develop systems that are able to learn from data and make decisions or predictions based on that learning, such as predicting the likelihood of a patient developing a particular disease. Computer vision could be used to develop systems that are able to recognize and classify objects or people in images or video, such as identifying suspicious activity in a surveillance video. Robotics could be used to develop robots that are able to perform a wide range of tasks, such as assisting with manufacturing or providing companionship to elderly individuals.

Overall, the future direction of AI research and development is likely to focus on a range of areas, including artificial general intelligence, AI safety, explainability, and more specialized areas such as natural language processing, machine learning, computer vision, and robotics. These advances are likely to have significant implications

www.ingramcontent.com/pod-product-compliance
Lightning Source LLC
Chambersburg PA
CBHW051912210526
45473CB00006B/1980